The Zoologist's Guide to the Galaxy

The Zoologist's Guide to the Galaxy

What Animals on Earth Reveal

About Aliens – and Ourselves

ARIK KERSHENBAUM

VIKING

an imprint of

PENGUIN BOOKS

VIKING

UK | USA | Canada | Ireland | Australia
India | New Zealand | South Africa

Viking is part of the Penguin Random House group of companies
whose addresses can be found at global.penguinrandomhouse.com

First published 2020
001

Copyright © Arik Kershenbaum, 2020

Set in 12/14.75 pt Bembo Book MT Std
Typeset by Jouve (UK), Milton Keynes
Printed and bound in Great Britain by Clays Ltd, Elcograf S.p.A.

A CIP catalogue record for this book is available from the British Library

ISBN: 978–0–241–40679–3

To my dog Darwin, who taught
me just how much different
species have in common.

And to my father, who taught
me to look for the differences
and the commonalities.

Contents

List of Illustrations

List of Illustrations

1. Introduction

That life exists elsewhere in the universe seems almost inevitable. That we could know anything about it seems almost impossible. But my goal is to show you that we can actually say a great deal about what aliens must be like, how they live, and how they behave.

We are increasingly confident that there is life elsewhere in the universe and, even more excitingly, that it might be possible to find it. In 2015, NASA's chief scientist, Ellen Stofan, predicted that we will have evidence of life on other planets within twenty to thirty years. Of course, she was talking about microbes, or their alien equivalent, not necessarily intelligent life. But the principle is still staggering. We have gone from being obsessed with the possibility of alien life at the beginning of the twentieth century, to complacent pessimism in the seventies and eighties, and now back to a realistic, scientific optimism. This book is about how we can use that realistic scientific approach to draw conclusions, with some confidence, about alien life – and intelligent alien life in particular.

Short of aliens landing in New York, how can we know what they're really like? Do we need to rely on the imagination of Hollywood, and of science fiction writers? Or perhaps alien animals are likely to be no more bizarre than a kangaroo hopping on giant feet, or a squid jet-propelling itself through the sea, with skin that flashes in rainbow colours. Trusting in the universal laws of biology that bind us – all life on Earth – and also those creatures on alien planets, we will see that the *reasons* for those adaptations that animals have adopted on Earth are likely to be reasons for adaptations on other planets as well. Hopping and squirting will be perfectly reasonable ways to get around on many different planets, just as they are on Earth.

How rare is life in the universe? Until the 1990s, the presence of planets around other stars (exoplanets) was a matter for speculation, and some mathematical calculations. We had no solid idea how many planets there might be in the galaxy, and what they would be like: how hot, how much gravity, what atmosphere, which chemicals? As technology reached a level where it was actually possible to detect planets around other stars, excitement began to grow. Perhaps there was indeed a possibility of detecting planets that could host alien life.

First indications were disappointing. The few planets discovered were large, hot and made of gas – not very amenable to life, either as we know it or otherwise. But less than twenty years after the first exoplanet was discovered, there was a major breakthrough. The Kepler space telescope was launched to search for possible planets; staring at stars in just one tiny fixed region of the sky. Within just six weeks of Kepler beginning operations, five new exoplanets had been discovered. By the time Kepler ceased operating in 2018, it had discovered an unbelievable 2,662 planets around the stars in just that tiny corner of the sky: about as much as you can cover with your fist at arm's length.

The implications are staggering. There are far more planets in the galaxy than we had previously thought, and with improved measurement methods we now know more about what those planets are like. We have found the full range of planetary conditions, from hot, Jupiter-sized gas planets, to those that are remarkably similar to Earth.★ The universe is now a lot more crowded than it appeared to be in 2009, and our grandchildren

---

★ See *The Planet Factory: Exoplanets and the Search for a Second Earth* by Elizabeth Tasker.

will likely not even believe that we once said, 'Earth-like planets are rare.' We no longer have any excuse to say that the universe lacks possible homes for alien life.

We now have a much better understanding of the physical environmental conditions that are likely to exist on alien planets and, increasingly often, we can even measure them directly. New instruments being developed will be able to detect the chemicals in a planet's atmosphere by spotting changes in the light passing through the atmosphere from the star it's orbiting. We'll be looking for oxygen, of course, but also for complex chemicals that might indicate industrial development. Ironically, pollution is a sign of cosmic intelligence.

Somehow, life came into existence at least once in the universe. We are proof of that. But how that happened, we don't know. There are certainly many theories on the mechanisms by which life may have arisen on Earth. Most likely, the basic chemicals needed for life formed at random and then, by a lucky chance, combined into a special kind of molecule that could make copies of itself. Overall, a fairly unlikely set of circumstances. Does that mean life on other planets arose the same way? Absolutely not. We really don't know how relevant the processes that we think took place on Earth may or may not be on other planets. Aliens may be based on carbon chemistry like ours, carbon chemistry unlike ours, or something altogether different.

The principles of chemistry are pretty well understood, so many of these ideas can be tested in the lab, seeing which chemicals might be stable and which not. We think that chemicals like the ones that make up our bodies are pretty good ingredients for something that is 'alive'. But beyond the most basic ideas of what alien biochemistry might be like, a thick fog descends. We have no examples of alien plants or alien animals to examine – nor even any idea whether the terms 'plants' and 'animals' would have any meaning on another planet. Despite NASA's optimism that we will discover signs of alien life, the vast distances between the stars means that it would take a huge technological leap to *visit* planets outside our solar system. We can mix alien chemicals in a lab, but watching alien birds with binoculars will be a much harder proposition.

One problem with understanding the nature of aliens is that our starting point for comparison is just one single type of life – that on Earth. How much can we use our single example of life to draw conclusions about other planets? Some people claim that speculation about the nature of alien life is futile; that our imagination is too tied to our own experience to be able to encompass the staggeringly diverse and unfamiliar possibilities that may be reality on other worlds. The science fiction writer and author of *2001: A Space Odyssey*, Arthur C. Clarke, said, 'Nowhere in space will we rest our eyes upon the familiar shapes of trees and plants, or any of the animals that share our world.' It's a popular belief that alien life is too alien to imagine. I don't agree. Science has given us the opportunity to move beyond such a pessimistic outlook, and we do seem to be able to identify some clues about what alien life may be like. This book is about using our understanding of how life works and, most importantly, how life *evolves*, to understand how life will be on other planets.

How did an Earth-bound zoologist like myself – who is more used to following wolves over the snows of the Rocky Mountains, or tracking furry hyraxes in the hills of Galilee – get involved in the search for extraterrestrial life? One of the things I study is animal communication, and why animals make the sounds that they do. In 2014, I gave a talk at the Radcliffe Institute in Harvard entitled, 'If birds could talk, would we notice?' It may seem obvious to us that humans have language and other animals do not – but how do we *know for sure* that is the case? I was looking for mathematical fingerprints of 'language' in the communication of animals – a clear-cut measurement that would say: 'Yes, this is language', or, 'No, this is not language'. With the encouragement of some good but somewhat eccentric colleagues, the obvious next step was to ask the same question about signals from outer space. Is this a language? If so, what kind of creatures might have produced it? From there, it becomes clear that we can extend our understanding of other aspects of life on Earth – finding food, reproducing, competing and co-operating with others – to alien planets as well.

But why study aliens in the context of *zoology*, when we haven't seen any aliens, and don't even know for sure whether they exist? When undergraduate students arrive at university, fresh from school and from exams that rigorously tested their ability to remember a long list of facts, our first job as their educators is to convince them that facts are all very well, but that they must understand concepts; not *what* happens in the natural world, but *why*. Understanding *processes* is the key to zoology on Earth – but it can also help us with understanding the zoology of other planets. As I write this, our second-year students here at Cambridge are preparing for a field trip to Borneo. Some of them are leaving the UK for the first time in their lives. Do we expect them to memorize a field guide to the hundreds of birds and thousands of insects of Borneo? Of course not. Like future explorers of an alien world, they must go equipped above all with an understanding of the evolutionary principles that led to the diversity of life that they will encounter. Only once the concepts are clear will interpreting the animals we find be possible.

Most people are confident that the laws of physics and chemistry are unequivocal and universal. They work here on Earth just as they would do on any exoplanet. Predictions that we make here about how physical and chemical materials will behave in different circumstances, are going to be good predictions about how those same materials would behave in those same circumstances in other parts of the universe. We rely on science to work in that way. Biology, however, is seen by some people to be an exception. We find it hard to believe that the laws we derive on Earth about how biology works would also apply on an exoplanet. Carl Sagan, one of the most famous astronomers of the twentieth century, and a fervent believer in intelligent life elsewhere in the universe, nonetheless wrote, 'For all we know, biology is literally mundane and provincial, and we may be familiar with but one special case in a universe of diverse biologies.'*

When dealing with the unknown, there are indeed good reasons for caution. But there are also reasons for optimism; we just need to be careful to choose those laws of biology that are truly universal, in the same way that the laws of physics are universal. Why should biology be 'mundane and provincial', rather than universal? Surely the laws of nature – physical, chemical *and* biological – are common across the universe? The Earth is unlikely to be so exceptional that the rules here are different to every other planet. Lucretius, the Roman philosopher who died in *c.*55 BCE, commented, 'Nature is not unique to the visible world.' Exoplanets have 'nature' as well, even if we've never seen them.

Contrary to what some people think, zoologists such as myself don't just spend our time identifying and classifying animals. Like scientists in all disciplines, we attempt to *explain* what we see in the world around us. Zoology, and evolutionary biology in general, is about proposing *mechanisms* for explaining the nature of life. *Why* do lions live in prides, but tigers hunt alone? *Why* do birds have only two wings? *Why*, for that matter, do the vast majority of animals have a left side and a right side? Observation is not enough. We want to derive a set of rules for life, in the same way that physicists derive rules for planets and stars. If

* See *Intelligent Life in the Universe* by I. S. Shklovskii and Carl Sagan, p.183.

those biological rules are universal rules, they will work as well on another planet as the law of gravity.

Yet there is no doubt that biology appears to be flighty and unpredictable. A physicist understands exactly how a ball rolls down a hill, and can give you a set of equations you can use to predict the motion of balls on hills everywhere in the universe. Physics experiments rely on highly controlled and simplified conditions – not at all what we find in the biological world. A well-known joke tells of a physicist trying to derive equations to predict the behaviour of a chicken, and declaring that this is possible, but only for a spherical chicken in a vacuum. Real chickens are out of the realm of 'physics' and so, a physicist would say, are unpredictable. But why can we predict the motion of a ball, and not the behaviour of a chicken?

Biological systems seem to avoid following strict rules because they are, in a profound way, complex. In mathematical terms, a complex system is one where multiple subsystems are mutually dependent on each other. It turns out that it doesn't take very much dependency between relatively simple systems for the overall behaviour to be utterly complex and unpredictable – *chaotic*, in the technical jargon. Imagine trying to predict the behaviour of all the interacting organs in your body. Better yet, imagine all the cells in all the organs – or all the proteins in all the cells in all the organs, and so on . . . The slightest change in one element can have a cascading, unpredictable effect. Even the simplest life is clearly complex. And complex systems are hard to predict.

One of the frustrating properties of an unpredictable complex or chaotic system is that no matter how hard you study it, you will never unlock all its secrets. We are used to the idea that if we investigate something carefully enough, we will come to understand everything about it. Science seems to be based on this idea. But chaos theory tells us that sometimes you can investigate a system a hundred times more carefully, and only gain ten times as much ability to predict what it's going to do. You can put more and more resources into understanding a complex system, but only get very marginal returns. That's clearly a mug's game. Fortunately, complex systems also have what are called *emergent* properties: you may not be able to predict exactly what they are going to do, but you can get the general idea. The chicken will look for seeds, even if I don't know exactly which seeds. In practice, being able to say 'the chicken will look for seeds' is *more* useful to me as a biologist than 'the chicken will look for *that* seed'. Rather than being able to predict how the biochemistry of alien life will work, or what their eyes will be made of, we can make general predictions that their biochemistry will provide them with energy, and also whether or not they will have eyes of any sort.

What then are these universal laws of biology, using which we can make confident predictions about life on other planets? The first and most important law is that complex life evolves by natural selection. It is hard to overemphasize the importance of this process, which has been the cornerstone of all biology since the seminal work of Charles Darwin. Natural selection is not just the only mechanism we know for creating complexity out of simplicity (if we reject the explanation of a divine force pushing complexity to develop), it is also an *inevitable* mechanism, not just restricted to the planet Earth, or to 'life as we know it'. If we see complexity in the universe – complexity of the kind that we would call 'life' – it is because natural selection has been operating.

Other excellent books have made the case for the universality of natural selection,★ but my claim is so extraordinary that in the next chapter I provide some extra detail on what I mean when I say that 'aliens evolved by natural selection'. As the

★ See *The Blind Watchmaker* by Richard Dawkins, notably.

philosopher Daniel Dennett pointed out, natural selection and intelligent design are almost the same thing: the accumulation of good features and the rejection of bad ones.* Whether we are designing an aeroplane or a paper clip, we keep hold of the good ideas from previous designs. Where selection and design differ, however, is that design has a goal in sight, whereas selection sees only one step at a time. A giraffe doesn't 'know' that it would be good to have a long neck, but it evolves one anyway.

This short-sightedness of natural selection actually makes our predictions about alien life much easier. We don't need to make grand predictions on how alien species 'should' be, we only need consider the conditions on a particular planet at a particular time to know what features are likely to arise. If we know a planet will have tall trees (or the equivalent), we can guess that some animals will have long necks, or long legs, or some equivalent.

Evolution by natural selection has another useful property: it is almost independent of the mechanism by which reproduction and selection take place. Famously, Richard Dawkins invented the term 'meme': a social concept or idea (like religion) which replicates through social communication, and competes with other ideas in a rather evolutionary way.† Natural selection can be defined in strict mathematical terms, without reference to any particular biological system or organic form of reproduction. For that reason, it is an incredibly powerful concept, and that simplicity and universality means that the concept of natural selection encompasses any likely route to complex life in the universe. Natural selection is not dependent on DNA, or any kind of Earth-bound biochemistry. So we don't need to

* *Darwin's Dangerous Idea: Evolution and the Meanings of Life* by Daniel C. Dennett.
† See *The Selfish Gene* by Richard Dawkins.

know exactly how alien biochemistry works – however it works, natural selection will be behind it.

Up until now the field of astrobiology, or the study of life outside of Earth, has traditionally focussed on a few clear areas. Mostly, astrobiologists study the origin of life: how life began on Earth and what that means for the possibility of life on other planets. Did life arise once on Earth, or many times? Did this miraculous event occur in warm, shallow lagoons, as Darwin suspected, or at underwater volcanoes, where the hot water and rich minerals create a perfect environment for weird and wonderful bacterial life?

Another important question is what alternative biochemistries might exist? Perhaps life on other planets does not use DNA for its genetic material, or maybe alien biochemistry is utterly different – based on a solvent other than water, for example. This is particularly important, as many planets (including some in our solar system) are too cold or too hot for liquid water to exist. However, these important fields are not the subject of this book. We want to investigate questions that astrobiologists rarely consider: what might *complex* alien life look like? Can we draw any concrete conclusions about the ecology and behaviour of alien life, using the tools and clues we have available on Earth?

A zoologist observing a newly discovered continent from afar would be buzzing with ideas about what kind of creatures might live there. Those ideas wouldn't be wild speculations, but keenly reasoned hypotheses based on the huge diversity of animals we already know, and how each animal's adaptations are well suited for the life they live: how they eat, sleep, find mates and build their dens. The more we know about how animals have adapted to the old world, the better we can speculate about the new.

This is the approach I'm going to use when talking about alien life – as different as it may be, there are still some things that we can learn from how life makes its living on Earth. The evolutionary processes that we observe here are due to pressures and mechanisms that are also very likely to occur elsewhere. Movement, communication, cooperation: these are evolutionary outcomes that are solutions to universal problems.

If we ever make contact with an alien civilization – intelligent beings, rather than just microbes or jellyfish – we can be confident of certain things: they will have a form of technology (or how could we contact them?), and that means that they are cooperative and therefore social. But merely knowing that a species is social unleashes an avalanche of additional evolutionary implications. They may be brutal and warlike, like us; but I will argue that to be social they must also be altruistic. An alien spaceship landing in central London would be a sure sign that the passengers 'spoke' to each other using some kind of language, but whether this speech might be acoustic, visual or even electrical, we cannot say. Two legs, many legs or none, I believe that it is language that will end up being our biggest feature in common with any alien civilization we encounter.

Rigorous scientific consideration of the possibilities of alien life are uncommon, but not unheard of. We are familiar with both the modern science fiction of *Star Trek* and the no more or less tenuous speculation of H. G. Wells's *War of the Worlds*. Ever since the planets were understood to be solid worlds of their own, attempts have been made to figure out whether or not there is life on them. In 1913, a British astronomer called Edward Walter Maunder★ published a short book, *Are the Planets Inhabited?* In it, he reviews with great scientific rigour the possibilities of life in

---

★ As well as this very readable book directed at the general public, Maunder was noted for his attempts to bring astronomy to the wider public. Together with his wife, Annie Russell, also an astronomer who graduated from Girton College, Cambridge at a time when women were not permitted to receive degrees, they founded the British Astronomical Association, in

our solar system, considering in turn each of the planets, the Moon and even the Sun (which was thought to be capable of supporting life by scientists as eminent as William Herschel, the discoverer of Uranus). One by one, he dismisses the possibility of life on Mercury, Mars, the Moon and the Sun, using clear reasoning based on the observations and measurements of the day. His reasoning can hardly be faulted, even by today's standards.

But his conclusions were often incorrect. Our understanding of the universe is limited, not only by our ability to reason logically but also by our ability to measure, and by our understanding of the mechanisms driving both the biological and physical processes that affect everything about the world around us. We may be making dire miscalculations, simply because some small feature of our knowledge is lacking. Maunder felt that Venus was the best candidate for life in the solar system, because the astronomers of the time estimated the surface temperature to be around 95°C and thought the thick clouds that cover the planet were made of water vapour. We now know, through better measuring equipment (not to mention spacecraft that have landed on Venus), that the surface temperature is nearer 450°C, and those beautiful bright white clouds are in fact made of sulphuric acid. Lack of good data will always hinder our search for explanations but, like Maunder, we cannot refuse to begin the search just because our data are imperfect.

We all want to know what aliens look like, but relying on the imagination of Hollywood producers is not going to be very realistic. Throughout the ages, people have imagined aliens either as exaggerated humans or exaggerated Earthling animals – perhaps giant spiders or worms, designed to induce nightmares.

---

protest against the Royal Astronomical Society, which did not permit women to become members.

The unknown and the dark are as frightening to us as they were to our ancestors before the invention of electric lights, and we fear that 'out there' may be animals and demons lying in wait. But despite its appeal on the big screen, lumping 'unknown' with 'scary' is not a rigorous way to pursue investigation. Can we be any more scientific about what aliens will look like? Unfortunately, the best efforts of serious reasoning still look slightly laughable – or are patently guesswork.

But how aliens *act* is much easier to predict than their appearance. Looks are more prone to accidents of evolution and quirks of embryological development; behaviour is a more fundamental response to the environment. We have two arms and two legs largely because of a bit of an evolutionary coincidence – our coelacanth-like ancestors used four fins to navigate the shallow waters in which they lived 400 million years ago. Those four limbs are still found in the descendants of that ancient fish: the amphibians, reptiles, birds and mammals we see today. With a different ancestor – some kind of crustacean, for instance – we could have had six legs, or eight. Whether or not we could have ended up with an odd number of legs, you must decide for yourself after Chapter Four – and see whether you also agree that aliens must have legs of some sort.

Behaviour serves a general purpose. For example, sociality (which we will look at in Chapter Seven) solves problems that exist on all worlds – problems you can't solve by yourself, like hunting animals larger than you, or building defensive structures in which to live. If aliens are faced with problems that they can't solve on their own, then some of those aliens are likely to be social. True, our social behaviour is generally idiosyncratic, and we don't necessarily expect aliens to possess religions like ours, or capitalist economies, but there are some features of sociality that must be universal. The mere existence of sociality depends on things like reciprocity, altruism and competition;

they drive the evolution of social behaviour, and would be present in any species that is, in fact, social.

Other chapters in this book deal with similar behavioural necessities, and what their evolutionary origins and implications might be: communication, intelligence and even language and culture have a role in shaping what we know of as humanity, and even these 'peculiarities' of human nature are not as idiosyncratic as they first appear. These features of our species could actually be a uniting similarity between us and aliens. Who cares if they are green or blue, so long as we both have families and pets, read and write books, and care for our children and our relatives?

Each chapter in this book is about some feature of animal behaviour on Earth that is not unique to Earth – that *can't* be unique to Earth. We have tended to invent strange-looking aliens, but we don't need to invent strangely *behaving* aliens, because the diversity right here on Earth already includes behaviours that are shared on other planets. In Chapter Two, I give an introduction to this idea – I explain why we are justified in using Earth-bound examples to understand life on other planets. Chapter Three looks at what it means to be an 'animal' – is that purely the definition of a creature on Earth, or could it also apply to those organisms utterly unrelated to anything on Earth? Chapters Four and Five talk about how animals, and aliens, move and how they communicate – two behaviours that we'd expect to find on any planet, and behaviours that are so constrained by the laws of physics that we can make a good guess at how they will work. Chapter Six is about that illusive (and prized) characteristic, intelligence: how animals make sense of the world around them and solve the problems they face. We all want to believe in intelligent aliens and, as I'll show in this chapter, it seems inevitable that they will, in fact, exist. Chapter Seven talks about another property we'd hope to find in aliens: cooperation and sociality.

So many animals on Earth live in groups, and for very good reasons – reasons that are not confined to our planet. Chapters Eight and Nine deal with exchanging information and with language itself, the one characteristic of humans that, so far, appears to be unique among life on Earth. Chapter Ten addresses the tricky question of artificial life, and whether alien planets would look very different if their inhabitants were not animals, as we know them, but robots or computers. Finally, Chapter Eleven attempts to address a difficult philosophical question: if intelligent, talking, social aliens exist, what does this say about the nature and uniqueness of humanity?

Our attempts to understand the nature of alien life are perhaps embryonic, but they have an important role to play in the development of astrobiology as a discipline, in the understanding of the science of life in general, and in the preparation for the time when humanity will have to come to terms with the fact that we are not alone in the universe. The question of how we will react as a species when we first discover that life exists on other planets is one that has not been sufficiently considered.* Will there be mass hysteria and looting? Religious fundamentalism, or a mass abandonment of religion? Or perhaps, according to the sixties hit 'Aquarius', 'Then peace will guide the planets, and love will steer the stars'? We cannot be ill served by being prepared.

The history of science is one of humans being knocked off their pedestal at the pinnacle of creation, and uncovering alien life will further emphasize our lack of uniqueness. Or will it? If evolutionary biologists such as myself are correct, our heritage is shared with all life in the universe. True, our origins are different, perhaps our biochemistries are very different, and we may not share a common ancestor with life on any other planet. But we share a common *process*. Our evolutionary history may not

---

* See *The Impact of Discovering Life Beyond Earth* by Steven J. Dick (ed.).

be identical to the inhabitants of other worlds, but at the very least we will be recognizable as intelligent life forms to alien zoologists on those alien planets.

If we live in cooperative societies and they do as well, then it is no small feat that we can identify the common evolutionary origin of our sociality. And maybe, just maybe, we will be able to use the term 'humanity' to mean something a little more broad, and more significant, than just the descendants of a group of apes that wandered across the grasslands of a tiny corner of one continent on a tiny planet in a corner of just one galaxy among billions.

2. Form vs Function:
What is Common Across Worlds?

In the early 1800s, the now legendary fossil hunters Mary and Joseph Anning uncovered an unusual skeleton on the beach at Lyme Regis, on the south coast of England. The scientists who examined it found it hard to classify; the bones looked like they belonged to a fish, and at the same time to a reptile. It was an ichthyosaur – a marine reptile superbly adapted to fast swimming, with a long snout and well-developed eyes. To most readers, that probably sounds a lot like a modern dolphin, but despite the apparent similarities between dolphins and ichthyosaurs, these two species are as unrelated as a human and a newt. Form – what an animal looks like and behaves like – is inextricably tied to function – how the animal makes its living, gathers energy and reproduces. This connection is the key to how we can know what *aliens* are like, without descending into fictional speculation.

I have promised that we will use the laws of biology – such as they are – just as we would use the laws of physics and chemistry to anchor ourselves in a set of fundamental, universal truths. If the nature of the universe is the same everywhere, then life conforms to the same rules everywhere. But just what are those biological laws? This is going to take some careful excavation. We want to be sure that we are not building a false world, full of speculative creatures, on the basis of our observations on Earth, inappropriately applied to other planets. It is a risky business to believe your own speculation, which, after all, exists only in your own mind.★

★ J. B. S. Haldane, in his essay 'Possible Worlds', wrote, 'Generally, philosophers who construct a funny world come to believe that it is the real world.'

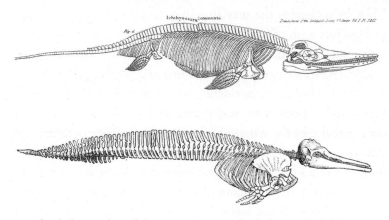

Top: the skeleton of an ichthyosaur. Bottom: the skeleton of a dolphin. Both animals seem to have lived similar lifestyles as fast underwater predators (function), and so evolved similar body structure (form).

Instead, we are on a search for those universal laws, the absolute basics of what constrains life, and dictates its nature. Remain sceptical. I may err. But our understanding of the biological nature of life, and in particular the evolution of life, has reached a point where, I believe, we can begin to generalize our understanding to other planets.

Our key tool for remaining anchored to reality, and not drifting away into science fiction, is to spot the crucial distinction between form and function. All animals have certain forms that are hugely diverse, and they impress us greatly. We admire the different colours of birds and flowers, the weird and wonderful shapes of elephant trunks and narwhal horns, and the howling of wolves and songs of humpback whales. The diversity of animal forms can be seen both in the way they look and in the way they act. The way they look in terms of their shape, size and colour, whether they're covered in fur or feathers, and whether they have trunks, tusks, shells, tentacles or any of the other myriad appendages that make different kinds of animals look so unique.

Their behaviour is the way that they look for food, find mates and interact with other animals of their own species and different ones. But each one of these forms, whether in appearance or behaviour, serves some purpose, some evolutionary role. Occasionally, evolutionary 'accidents' do occur, where a form has no function, but persists anyway – perhaps the particular form was once useful, like the wings of an ostrich, but no longer serves any purpose, and there's no evolutionary reason to change. But most forms serve a function: birds are colourful to attract mates; elephants' trunks are for manipulating food and other important objects. Almost all the forms we see provide some function that improves the animal's ability to live, thrive and survive – even if it often isn't apparent what that advantage might be.

Why do zebras have stripes? Scientists argued for many years over the possible reasons. Charles Darwin himself doubted the oft-quoted explanation that the stripes act as camouflage, and various alternatives have been proposed: to signal quality to members of the opposite sex, to confuse predators with psychedelic patterns of moving lines, to confuse biting flies trying to land on the animals, or even to help the animals keep cool by generating breezes between the warmer dark stripes and the cooler white ones. The important point is not who was right and who was wrong. The key point is that every explanation must propose *some* advantage, some *function*. The familiar forms of life that we see around us have evolved because they provide specific advantages.

Despite this, sometimes random events can seal the future fate of life forms, without providing a specific function. This is especially important when populations are made of very few individuals. If the future human colonists of an alien planet, or the first birds arriving on a distant island, are all very similar genetically, their offspring will reflect that lack of diversity for generations to come. When populations are isolated, random changes that are neither useful nor harmful accumulate, and cause different species to look distinct. We do have to be careful when observing new species – whether on other planets, or on long-lost islands on Earth – not to assume that all forms are directly attributable to a particular function. This is called neutral selection, and there's considerable debate over how important it is for evolution. However, that kind of coincidental form is usually mundane, modest, certainly not dramatic and ultimately costly – as zebra stripes probably are, in making the animals more visible to predators.

The distinction between form and function is the most important step to take in divesting ourselves of our imagined alien life forms. Mostly, these fictional aliens take a shape similar to those built in Hollywood for screen and television, and these in turn are little more than exaggerated humans, with physical characteristics emphasized (e.g. teeth, eyes) to invoke a sense of exaggerated human abstract characteristics (e.g. voraciousness, intelligence). But this book is just as much about what aliens *act* like, as what they *look* like. Those laws of biology that we share with alien life forms tend to constrain the way that they solve the challenging problems of life: finding food, avoiding becoming food, and reproducing.

What initially impresses us most about life on Earth is its forms, rather than its functions: the bright colours of birds, flowers and poison dart frogs; the sheer size of a blue whale; the tenacity of a lion bringing down a buffalo. But when we stop to think about it, this diversity of forms is really only a reflection of the diversity of functions. Animals are diverse because they have to solve a diverse set of problems: colourful to attract mates or warn off potential predators; large as physical protection against those predators; tenacious to get food. Our very general and universal laws of biology may not allow us to make very *specific* predictions about the forms of life on other planets, but we will be able to make general predications about the functions that these animals fulfil, and we can be sure that within those general functions, there will be as great a diversity of form as there is on Earth. If the equivalent of birds live on other planets, then the birds will have different colours just like on Earth. We just don't know what those colours will be, or even if they will be 'colours' as you and I perceive them.

However, in case you are disappointed that you will not get to know from this book whether or not aliens are, in fact, green, there is a considerable upside to examining function, before considering form. The way that aliens adapt to their environment and the challenges it throws them, is ultimately more interesting than the way they look. At the very least, these behavioural adaptations are more likely to be shared between them and us, and we will likely have more in common with intelligent aliens in our behaviour than in our appearance. In this chapter, I will hope to persuade you of this distinction between form and function, and show why function is so much more vital. To do this, we will have to revisit some of the principles of natural selection and evolution, as well as why these principles are going to be shared on other planets as well as on Earth.

Natural selection: the universal mechanism

Explaining how complex life forms exist at all is a far harder challenge than it may seem at first glance. Complex life exists in the face of the most relentless of the laws of physics:* that order tends to disorder, complexity to simplicity, information into nonsense. As a drop of ink spreads throughout a glass of water,

---

* In particular, the First Law of Thermodynamics: energy cannot be created or destroyed ('There's no such thing as a free lunch'); and the Second Law of Thermodynamics: useful energy always decreases ('You can't even break even').

buildings crumble and flesh decomposes. Providing a definition of life has been a challenge for philosophers for as long as humans have existed, but any definition must surely include this property of fighting the universal tendency to disorder: not crumbling, not decomposing, not dying. Given that a boulder will always tend to roll down a hill, how can we get a boulder to the top of the hill? Given that the universe appears to be predisposed against life, how can life exist anyway? We need a mechanistic explanation, one that explains, step by step, how life can not only exist, but become more and more complex – surely the opposite of those relentless laws of physics!

Let us first discount the idea that a fully fledged complex life form popped into existence – that is just too unlikely, without an even more complex life form to create it. Perhaps a deity really did create the universe fully formed, but in that case we could say absolutely nothing about alien life. All the shapes and colours and behaviours of alien creatures would be no more than the whim of their creator. Stephen Hawking said that we could, in fact, know the mind of God, but only by understanding all the laws of physics in their entirety.★ We are quite a long way from that goal.

So life begins with something simple. How can such a simple life form become more complex? Does it *know* what extra complexity it wants to gain? Although we can consider that a human might decide that a bionic arm would be a good idea, it's hard to see that a primitive cell or molecule would be gifted with such foresight (we will discuss this more in Chapter Ten). We are searching for a 'good' explanation of the complexity of life: and to be a good explanation it must be self-contained, without appealing to any external, undefined processes (like a god), or processes that we don't believe are possible (like a molecule that

---

★ *A Brief History of Time: From the Big Bang to Black Holes* by Stephen Hawking.

'knows' what it wants to become). Complexity must accumulate by itself, and so a key component of a good explanation *must* be that it doesn't require any foresight, otherwise we wouldn't be able to apply it to the earliest, simplest life forms.

Even if we accept that we don't know how an original life form was created, we are forced to explain how it can become *more* complex. I, together with almost every other modern scientist, claim that natural selection appears to be universal, and the ubiquitous way of explaining the fact that life is more complex now than when it was created 3.5 billion years ago. But what is this 'natural selection', and why should it be the universal explanation for complex life?

At its simplest level, natural selection is easy to understand. Beneficial traits *accumulate*. Some new features will survive, other innovations will not, but good ideas developed by previous generations are not forgotten. Richard Dawkins explained this process with beautiful simplicity in his book *The Blind Watchmaker*. Imagine randomly choosing a string of twenty letters, say SDFLKJFGOSDIFHGSOFGH. The chances of arriving at a particular sequence, say, 'The Blind Watchmaker', are astronomically small: actually, one in 42 billion billion billion.[*] No one believes that order can arise out of chaos randomly. But if each time you make some random changes to the sequence above, you keep changes that match the sequence we're looking for, 'The Blind Watchmaker', the result is completely different. Good innovations – say, changing the initial 'S' (which isn't in the target sequence) for a 'T' (which is the first letter of 'The') – don't disappear, so bit by bit the best sequence, i.e. the 'right' sequence, will emerge. Remarkably, using this 'selection' approach, the

---

[*] For a sequence of twenty characters, with twenty-six possible letters plus the 'space' character, the number of possible combinations is $27^{20} = 42,391, 158,275,216,203,514,294,433,201$.

correct sequence emerges after about just 540 attempts — an improvement by a factor of about 80 million billion billion!*

Of course, in nature there is no foresight. There is no 'right' sequence. But there are better sequences and worse sequences. As long as good changes accumulate, our sequence gets better and better. We *can* push a boulder up a hill, if your hill is actually made of steps, and you can take a break at each step. Take one step at a time, each time waiting until you chance upon a way to get up the next step. *That* is the core of natural selection, and it is beautifully simple and obvious.

But are there any alternative explanations?

Natural selection has the rather remarkable property that scientists are at a loss when it comes to proposing realistic alternatives. Usually, when an explanation for a natural phenomenon is in doubt, a number of different alternatives are weighed up against each other, and the most convincing one is tentatively accepted (until further evidence causes us to change our minds). 'Light' could be something that streams out of visible objects, or it could be a sensory beam sweeping out of our eyes (as some ancient Greek philosophers believed). Both are reasonable hypotheses — until suitable experiments are performed to distinguish between them. For much of the Classical period, the ideas that the Earth was flat and that the Earth was spherical coexisted, with various supporters and opponents on both sides, until Eratosthenes performed his brilliant experiment to measure the radius of the (very much spherical) Earth in 240 BCE.

In the case of natural selection, however, it is remarkable that apart from a few deeply unsatisfying and non-scientific

---

* The average time to 'correct' a letter is twenty-seven attempts (as the correct letter is one among twenty-seven), so the total expected number of attempts is $27 \times 20 = 540$.

explanations for the existence of complex life, there are no serious alternative contenders.

Perhaps we need to think harder; maybe we're not being clever enough. 'That's the only answer I can think of' is hardly a rigorous explanation. However, although a lack of alternatives is not proof, it is at the very least a sign that natural selection is a likely candidate. All the competing ideas that have been proposed to explain the origin of complex life are very much descriptive, rather than explanatory.

To start with, there is the possibility that an all-powerful deity 'directs' individual changes in the form and behaviour of creatures, nudging them along the evolutionary road. Alternatively, there may be some as yet undiscovered 'life force' that drives species changes. Or perhaps life is created with the template of all future developments already inside it – a kind of blueprint for humans, waiting inside a bacterium. All that's needed is to peel away the layers and voila: us. But these are all descriptions, rather than *explanations* of how complexity arises. Every human culture has some kind of creation story, none of which can be weighed up against the others in any objective way. Stories do not provide us with explanations, and we have a strong desire to find a *mechanism*, not just a story.

Mathematical analyses can give us a strong indication that natural selection may be the only explanation for life in the universe, and much of our understanding of the inevitability of natural selection as the *mechanism* by which life evolved is mathematical in nature. The equations are perhaps a little dry, but the ideas behind them are most definitely not. One of the most complete mathematical descriptions of how and why evolution occurs was put together by one of the most remarkable, and unknown, characters of twentieth-century science. George Price was a chemist – not a biologist or a mathematician – but he collaborated with two of the giants of evolutionary theory, John

Maynard Smith* and Bill Hamilton, to build the most complete mathematical description of why evolution happens. According to popular accounts, Price was so taken by the inevitability of evolutionary forces that despite having been a committed atheist, he converted to Christianity, gave away his possessions, devoted the rest of his life to helping the homeless and sank into a deep depression, eventually dying in a dilapidated squat.†

One of the most important elements of Price's equation is that both the measure of an animal's trait – say, how long its teeth are – and the advantage given by a particular length of tooth *vary*. Some animals have longer teeth, and some shorter. While it might well be an advantage to have longer teeth, it doesn't always follow that an animal with teeth twice as long would have twice the advantage. Rather, there is a *tendency* to greater advantage with longer teeth. Price showed mathematically that the rate at which a trait will change over time in a population – the rate at which the descendants of animals get longer and longer teeth – depends on the 'covariance' between the trait and the advantage it provides, in other words, how tightly tied together are the trait and its advantage. If doubling the length of the teeth always gives double the advantage, that long tooth will spread through the population like wildfire. If the link is more tenuous – say, if a tooth twice as long gives a 10 per cent advantage, but only 50 per cent of the time – then the pace of evolution is slower.

It cannot be overstated how crucial it was that scientists were now armed with a model that predicted how evolution progresses. And this model makes no assumptions that are tied to the planet Earth. Price's equation works just as well on every exoplanet in the galaxy. As the British philosopher Bertrand Russell said, 'I like

---

* See *The Theory of Evolution* by John Maynard Smith.
† *The Price of Altruism: George Price and the Search for the Origins of Kindness* by Oren Harman.

mathematics largely because it is not human and has nothing particular to do with this planet or with the whole accidental universe – because, like Spinoza's God, it won't love us in return.'

We can make some of the mathematics more visually intuitive. Imagine that you are dropped off in the middle of mountainous countryside in thick fog and told to find your way to the top of the mountain. The landscape around you is called a 'fitness landscape'. Although easily confused with the cardiovascular fitness you might gain by climbing mountains, this is something completely different. Evolutionary 'fitness' can be thought of as how effectively you project your own genes into future generations. Not just how well you survive, but how many offspring you have, and how well they in turn survive to have offspring, and so on and so on, over the generations. In the case of our visual analogy, the higher your altitude, the better adapted you are to your environment, and the more evolutionary fitness you have; the higher up a mountain you are, the more offspring you raise successfully. How many different methods could you use to find a mountain peak? Perhaps you have a map or perhaps you can see the top of the mountain and head towards it. But otherwise the only way to find your way is to look around you and see which direction is uphill and always to follow the uphill slope. Now if I told you not to follow the uphill path but to think of some alternative way of reaching the top of the mountain, would you be able to? There really *isn't* any other way. You could try jumping around at random, but this can be mathematically proven to be ineffective. Small steps, local improvements, are the only tool available. And that's natural selection.

Of course, science is always open to new discoveries and radical new ideas that knock down the firm foundations on which we thought we stood. No one would mind if an alternative to natural selection were discovered. But that is *not* the same as saying there 'might' be an alternative. Admitting that our understanding of physics is incomplete is not the same as saying 'maybe there are ghosts and fairies so we can reject quantum physics'. You can always propose empty 'maybes', but they aren't particularly useful.

The famous British astronomer Fred Hoyle was a tremendously important character in the development of twentieth-century astronomy, but he also wrote excellent science fiction. His 1957 book *The Black Cloud* was not just a brilliant and convincing story of what alien life might be like, but also an accurate description of the way that scientists approach dealing with the unknown. In it, he postulated a giant cloud of gas, hundreds of thousands of kilometres across, that was sentient and highly intelligent. His description of how such an alien could exist and function were exemplary, but he has been criticized for his lack of biological insight – he didn't explain how such a creature could *evolve*! What were the steps that led to the existence of such a hyper-intelligent cloud of gas? What was the precursor of such a cloud like, and how did it become different: how did it become the contemporary cloud?

Such an oversight is very common when fantasizing about aliens – they may be intelligent, possessing incredible abilities such as telepathy, telekinesis or the power to alter reality with a click of their fingers. But *why*? Why would such an unlikely state of affairs arise? The only answer is as an improvement on a previous situation. Pushing a boulder up a hill one step at a time. Natural selection.

As it happens, Hoyle had a simple answer to this particular criticism of his story. At the time, in the 1950s, a debate was raging about why all the galaxies in the universe appeared to be moving away from us. Two theories were proposed: that in the beginning the universe was very small, and has been expanding ever since, or that the universe had no beginning, and has always been expanding, with new matter being created at the centre all the time. Hoyle thought that the former explanation was nonsense, and mockingly dubbed it 'the Big Bang'. The name stuck, and of course we now accept this to be the correct explanation. But at the time, and on the basis of the existing observations, Hoyle and others insisted that the universe had no beginning. So it is no surprise that when his fictional scientists ask the Black Cloud how the first of its species arose, the Cloud replies, 'I would not agree that there ever was a first,' and the gleeful scientists react: 'That's one in the eye for the exploding-universe boys!'

If time is unlimited, then we must rethink the nature of the origin of life. But the scientific community is confident that the universe did indeed have a beginning, and so life too must have had a finite beginning – and life must have developed and diversified from that beginning. Natural selection is our universal explanation for this process.

Convergence: our key to alien life

My bold claim that we can apply lessons from studying life on Earth to life elsewhere in the universe comes from a simple observation: evolution seems to work similarly in similar environments. Both birds and bats fly, yet the common ancestor of the birds and the bats lived 320 million years ago, long before the dinosaurs, when reptiles were just beginning to take over the world. They certainly didn't fly, because the descendants of that reptile ancestor include not just birds and bats but all snakes and turtles, dinosaurs and mammals, from elephants to humans. Clearly, the ability to fly evolved separately in birds and bats at some later stage.

In fact, we know that powered flight evolved at least four

times on Earth. Birds evolved flight around 150 million years ago, when dinosaurs roamed the Earth. The famous fossil *Archaeopteryx*, dating from around this time, appears to be halfway between a dinosaur and a bird and caused much consternation and scratching of heads among scientists in the nineteenth century – including Charles Darwin. Bats, on the other hand, didn't evolve to fly until a little more than 50 million years ago, almost certainly after the extinction of the dinosaurs. Bird wings and bat wings are so strikingly different that it is hard to believe they perform similar functions. Bats have tremendously elongated fingers, which stretch to the rear edge of each wing, with a thin membrane in between, rather like the webbed feet of ducks, except that the webbing extends all the way up the arm as well. Birds, of course, have feathers rather than skin covering their wing, but unlike bats, their bones only run along the very leading edge of the wing, with feathers trailing behind them.

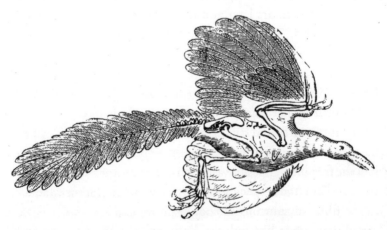

An artist's impression of *Archaeopteryx* from 1871 in *On the Genesis of Species*, by St George Jackson Mivart, a contemporary and correspondent of Charles Darwin. Like modern birds, the feathers trail behind the wing bones, which support only the front of the wing.

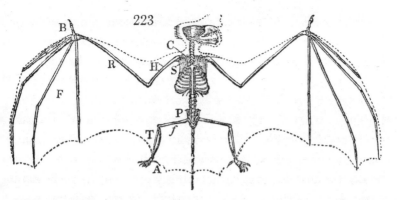

A drawing of a bat skeleton from *Animal and Vegetable Physiology, Considered with Reference to Natural Theology*, by Peter Mark Roget in 1834. The finger bones extend right to the back of the wing, supporting the membrane.

Despite the fact that the flight of birds and bats evolved quite separately, the uses to which flight has been put are very similar. If you watch swallows and swifts darting through the air to catch flying insects on the wing, they look remarkably like the bats you'd see a few hours later at dusk, flitting around after insects, just like a bird. The tiny Nathusius' pipistrelle bat, weighing about 10g, manages to migrate hundreds or even thousands of kilometres, rivalling the long-distance journeys of many birds.★ Whatever its origin, flight is an incredibly useful function, and it isn't surprising that it crops up again and again.

Of course, birds and bats aren't the only flying animals. Pterosaurs, enormous flying reptiles, took to the air even longer ago than the first birds, possibly as much as 220 million years ago. For some of them (those species immortalized in numerous caveman horror films of spurious biological accuracy) their huge wings helped them glide like vultures, but exactly how they managed

★ http://bats.org.uk is a website for citizen science pipistrelle tracking.

to take to the air remains under intense investigation.* But there's no doubt that they evolved quite separately from birds: pterosaurs weren't dinosaurs, whereas birds descended directly from fast-footed dinosaurs, which were more closely related to *Tyrannosaurus rex*.

The fourth instance of the evolution of flight – and the most widespread example of flight on this planet – goes even further back, to the dawn of the insects 350 million years ago. When insects became the first really successful land animals, their rapid evolution gave rise to a huge diversity of forms, including some unique adaptations to life in their new environment. Unlike the slow and gentle sinking of life in the oceans, if you fall off a tree, you hit the ground fast! Perhaps early wings provided the ability to slow one's descent, and even to steer the falling animal back to the tree trunk, to avoid the tedious climb back up from the ground (a technique still used today by flying squirrels, gliding from tree to tree with the help of a flap of skin between their front and rear legs).

The sheer utility of flight to these small creatures that swarmed across the newly inhabited land led to a vast diversity of flying solutions: irritating buzzing mosquitoes, graceful dragonflies, unlikely-looking flying beetles and, of course, the bumbling bee, whose flight looks almost impossible. It is difficult to doubt that insect flight and bat flight are separate mechanisms, evolved separately, but it should be clear that flight itself is unbelievably useful.

The way that similar solutions – in this case, flight – have evolved separately in distantly related species is part of a phenomenon called convergent evolution.† Presented with similar

---

* See *Restless Creatures: The Story of Life in Ten Movements* by Matt Wilkinson.
† Strictly speaking, we are more interested in 'parallel' evolution, rather than 'convergent' evolution. In the latter, we assume some kind of family

environmental challenges, similar solutions seem advantageous. Indeed, it is quite likely that given a particular problem, there exist only a limited number of *possible* solutions. If that is the case, it should not surprise us that birds, bats, pterosaurs and insects have arrived at similar functions, albeit with different forms.

This example of the convergent evolution of flight only scratches the surface of a hugely broad phenomenon. Convergence is everywhere. Eyes like ours with a large lens evolved at least six times. The generation of an electric field from the body (either to stun prey or to sense surroundings) has evolved at least as many times. Giving birth to live young appears to have evolved (quite independently) more than 100 times. Even photosynthesis, the basis of all life on Earth, probably evolved separately in at least thirty-one lineages.*

Perhaps the most famous example of convergent evolution is that of the recently extinct marsupial predator known as a thylacine, or a Tasmanian wolf. The last known thylacine died in a zoo in 1936, but they were widespread across Australia and New Guinea until the arrival of humans and dingoes several thousand years ago. The similarity between the thylacine and the dog-like canids, including wolves and coyotes, is uncanny; you might easily mistake it for a peculiar breed of dog. However, the thylacine was a marsupial with a pouch like a kangaroo or a koala bear, and no more related to a wolf than it is to a bat. How could such physical similarities evolve in such unrelated species? The answer (as of course will be familiar to you by now): both

---

relationship – any kind – between species. As aliens are not likely to be related to life on Earth at all, we should use the more general term of parallel evolution. However, I think 'convergent' is more evocative, and hope the reader will indulge this poetic licence.

* Simon Conway Morris's book *Life's Solution: Inevitable Humans in a Lonely Universe* lists these and many other cases of convergent evolution in great detail.

the thylacine and its dog-like counterparts evolved to take advantage of similar ecological niches.

Now lost forever, it is hard to know exactly how the thylacine hunted. Did it chase down bounding kangaroos in packs, much like modern-day wolves chase caribou? Or did it pounce on unsuspecting prey, much like the probable ancestor of our domestic dogs, perhaps using its peculiar back stripes as camouflage? Simply by asking these questions, we have cemented our certainty of convergent evolution. We can study skeletons to determine how strong was the bite of the thylacine jaw (researchers have determined that it was not very strong), and how suitable the elbow joint was for long-distance pursuit (similarly, not very), and conclude that the thylacine was more likely an ambush

A photograph of two thylacines in the National Zoo in Washington DC, from the Smithsonian Report of 1904.

predator than a coursing predator like a wolf. But such an analysis shows how powerful an assumption of convergence we have made: similar ecologies produce similar features.

And here is the main point. Convergent evolution is *not* just a phenomenon restricted to life on Earth. The same principles that lead birds and bats to evolve similar solutions will also lead alien birds and bats to fly. There is nothing special about Earth or about the (very distant) family relationship between birds and bats. Convergence between species on different planets that occupy similar niches is almost certain to occur.

It may sound at this point as if I am proposing that alien planets (or at least those that are physically similar to Earth) will be teeming with Earth-like creatures: alien wolves and bats, alien kangaroos and blue whales. If wolves and thylacines evolved separately but in parallel when completely isolated geographically, why should that not be true for all life? What if the very first form of life on Earth, whatever it was, perhaps the first ball of proteins and RNA wrapped up in a fatty bubble, also coincidentally was the first form of life on another planet? Would that planet also evolve four-legged wolves, six-legged beetles and two-legged humans?

But there are also reasons to think that such convergence may not be as widespread as we might think. The ecologist Stephen J. Gould famously proposed a thought experiment in which we rewind the 'tape of life' to a suitably early point, and then press 'play'.* Would we expect, after billions of years' playback, to end up right back where we are now, with exactly the same species and exactly the same evolutionary history? Probably not. The long story of life on Earth has been one of constant evolution, yes, but also one of repeated disasters and lucky

---

* See *Wonderful Life: The Burgess Shale and the Nature of History* by Stephen J. Gould.

escapes. Not long after complex life arose, the entire Earth froze from pole to pole, an event dubbed 'snowball Earth'. Luckily, some organisms survived in the still-liquid oceans under the thick ice. When an asteroid the size of Cambridge, England, hit the Earth 66 million years ago, all large land animals went extinct, and with the dinosaurs gone from their niches, small mammals rapidly took their place and became the horses, tigers and armadillos of today. Surely if that asteroid had been just a few hundred kilometres off course, it would have missed the Earth completely, and the last 60 million years of evolution would have looked very different. Can we really be confident about what life will evolve on a planet, when so much seems to be at the mercy of seemingly random astronomical events?

Even the asteroid that led to the demise of the dinosaurs pales into insignificance compared to the unimaginable mass extinction known as the Permian–Triassic event that occurred 250 million years ago. We don't know why, but a sudden shift in the chemical balance of the atmosphere and the seas occurred, causing such a huge mass extinction that life was almost wiped out completely.* Perhaps 90 per cent of all species became extinct. There can be no doubt at all that, even given the power of convergent evolution, life on Earth has been massively shaped by these unpredictable disasters, and it seems likely that other planets have also experienced their share of lucky misses (or, perhaps, they have mostly not been so lucky, and no one survived). How can we say that we know what life will act like on other planets, when the tape of life could not even be replayed reliably here on Earth?

The mass extinction at the end of the Permian period gives us a clue to the answer to this question. Life barely survived. But within 10 million years, it was back up and running again

---

* *When Life Nearly Died: The Greatest Mass Extinction of All Time* by Michael J. Benton.

(although it would be many times longer before life was once more as diverse on Earth as it was during the Permian period). True, almost all the dominant species from before that time were gone, including the iconic trilobites that used to scuttle across the ocean floor like modern crabs. But the niches remained. As long as there was food on the ocean floor, something could take advantage of that food by scuttling across it. And that's exactly what happened. The rise of both mammals and dinosaurs began in that empty age of recovery from the mass extinction. For millions of years there were only a few hardy species that survived in those harsh conditions – think of weeds and rats that rapidly take advantage of an abandoned industrial estate. But there was sunlight, so there were plants, and so there were also other animals to eat. When you wake up to a world almost empty of competition, your opportunities suddenly seem endless.

Life exploded forward in a vast range of new forms. Evolution went into overdrive, simply because so many empty niches were available. Species that were highly specialized, most closely tied to their niche and to some particular conditions or particular food source, were the most susceptible to extinction. Those species that were a little more flexible, a little more able to exploit new opportunities, may have survived and then flourished on those opportunities. In what evolutionary biologists call an adaptive radiation, surviving populations began to specialize in exploiting the now empty niches and evolving their own specialist adaptations to do so. The imagery is apt: species radiating out into new habitats and ecologies, adapting as they go, and diversifying both into new functions and into new forms. As in the story of the Three Little Pigs, suddenly different opportunities existed for building houses out of different available materials, and every opportunity was exploited. Adaptive radiations are thought to be absolutely essential for a biosphere as rich and diverse as the one we have now, and so catastrophes (although hopefully no more disastrous than the Permian mass extinction) are an essential part of the diversity of life.

So, when life returned to its full array, many of the old niches still existed and were filled by new occupants. If trilobites were no more, crustaceans could do their job of scouring the ocean floor for food. The forms were radically different (crabs instead of trilobites), but many of the *functions* of how animals found food and protected themselves against predators persisted. Plants were widespread on land before the mass extinction, and so were herbivores grazing them, and the predators that fed on the herbivores. Many of these herbivorous and predatory animals like *Dimetrodon* looked like giant reptiles, but were actually the ancestors of mammals. After the extinction, the giant reptiles appeared and came to dominate the world in the form of the dinosaurs. But when you look at a *Dimetrodon* from before the Permian extinction, and dinosaurs like *Agathaumas* from 200 million years later, you cannot help feeling that the tape of life, although not replaying the forms of the previous showing, is at least duplicating many of the functions.

Artistic impressions by the natural history illustrator Charles R. Knight (1874–1953) of a *Dimetrodon* (left) and an *Agathaumas* (right). Despite similarities, the *Agathaumas* is a dinosaur, a reptile with legs directly under its body, whereas *Dimetrodon* is not even a true reptile, and related to the ancestors of mammals.

The laws of sex

The distinction between form and function, and the way that evolution fills the different opportunities (functional niches) with multitudinous forms, is the first step towards applying universal 'biological laws'. But what about the laws themselves? To what extent is our understanding of natural selection universally applicable, and to what extent is it dependent on the peculiar conditions that we have observed on our planet? Natural selection in its most basic form is universal and assured. Although it has been popularized by the phrase 'the survival of the fittest', the core of natural selection is (only slightly) more sophisticated.

Natural selection occurs whenever individuals inherit characteristics from their parents, those characteristics vary throughout the population, and different characteristics give rise to different 'fitness' in individuals – where 'fitness' means your ability to be represented by your offspring in future generations. So 'survival' is usually good, because you can have lots of offspring, but if you live fast and reproduce prolifically, that's also good. Caring for your offspring might also be good for your fitness, because then they themselves are likely to survive and have more offspring – and so on and so on. It's a very likely (although perhaps not universally accepted) assumption that *any* system showing these properties – heritable variation and differential fitness – will experience natural selection. That is, natural selection is *inevitable*, even in non-biological systems like computer programs, internet memes, religious beliefs and so on, but especially in biological systems. There seems no doubt that organisms on other planets experience natural selection in this broad sense, because there is no other mechanism we know that can spontaneously produce and maintain the kind of complexity that we call life.

However, in the 150 years since *On the Origin of Species* was published, evolutionary science has advanced, and we now know more – much more – about the mechanisms that have driven the complexity of life on Earth. These mechanisms, which we will discuss below, are essentially variations on the principles of natural selection, but their universality is far more difficult to assure. We have excellent and complex mathematical models of the way that evolution works, because there's a lot more to explaining the diversity of animals and plants on this planet than just the survival of the fittest.★ One of the reasons why the Price equation was so important is that it incorporated other vital elements of natural selection: like how related one animal is to another. But while these models represent a triumph of scientific investigation, a nagging doubt remains. We build models based on our experiences, and our implicit assumptions that life is what we see around us. Perhaps those assumptions contain, buried within them, certain peculiarities of life *on Earth only*, derived from our own investigations and our own experiences here? Is it possible that while natural selection is the driving force of evolution around the universe, the details may be different – possibly very different – from the way that evolution works here?

Two of these supplemental mechanisms are particularly concerning: sex and family. Many of the most impressive animal forms and behaviours that we see around us are somehow connected to attracting a mate: bird colours and birdsong, deer antlers, the elaborate mating displays of male wolf spiders (drumming their legs and slamming their bodies on the ground), male bighorn sheep colliding head-on with each other in a fight for dominance. And, of course, the iconic tail-feather display of the peacock. All these traits have evolved because they improve the chances of mating, even if they seem to be highly problematic (to

★ *The Blind Watchmaker* by Richard Dawkins is an excellent introduction to the kinds of processes that have been driving evolution on Earth.

say the least) in terms of survival. Peacocks can be caught by their long tails, bighorn sheep often fracture their skulls in their collisions, and colourful male birds are less camouflaged than their female partners. It's easy to marvel at the seemingly wasteful behaviour of animals when you watch a skylark hovering above his territory singing at the top of his voice – an activity that is hugely demanding on his energy reserves. Natural selection is still occurring, of course, but the emphasis here is on competing for offspring, rather than competing for survival, and this process is called sexual selection.

The other mechanism that is of concern is a variant on natural selection where animals give aid to other animals to whom they are related: kin selection. This can be as simple as parents providing care for their offspring, or as complex as meerkat gangs in which subordinate females nurse the babies of the dominant female (often their sister). That means that they are sacrificing both their own energy reserves and their opportunity to have babies of their own, to provide care to their nieces and nephews. Kin selection is particularly important because it is largely responsible for the evolution of social behaviour as a whole (which we will discuss in detail in Chapter Seven), and social behaviour is probably necessary for the development of any kind of space programme among the alien species we hope to meet.

It probably hasn't escaped your notice that both kin selection and sexual selection rely on one peculiar property of life on Earth. Sex. Those spectacular peacock feathers would not evolve without sex, and complex social behaviour would not evolve without sex. So do aliens have sex?

I wish there were an easy way to answer this question, but unfortunately we know surprisingly little about the origin of sex on Earth, or even why it exists at all, which makes it very difficult to speculate about what might happen on other planets.* Non-sexual (asexual) reproduction means producing clones, offspring that are identical to you. Occasionally, an error in the genetic mechanism occurs, and some diversity can arise that way, but basically all of your family will pretty much be identical. In contrast, sexual reproduction means shuffling your genes with your mate, and your offspring are much more diverse; some look more like you, some less. So sexual reproduction seems hugely inefficient – not only does it take a lot of time and energy, but your children only have half of your genes. Nonetheless, sexually reproducing animals (and plants) are ubiquitous on Earth, and are more diverse in both their form and function than the rather uncomplicated asexual bacteria. Strawberry plants mostly reproduce asexually, sending out runners that spawn new plants, clones of the original, and rarely do daughter plants sprout from seeds. But the delicious strawberry fruits (seeds and all) are the product of sexual reproduction.

Many theories exist to explain the origin and persistence of sexual reproduction – to avoid the unchecked spread of parasites, for example, by ensuring that each generation has enough genetic innovation to keep one step ahead of the pathogens – but whatever the origins, we can observe a couple of points retrospectively. Firstly, sexual reproduction seems to have been absolutely crucial to the evolution of life on Earth, although the reasons why aren't clear. Possibly evolution is faster with sex, and possibly it is more reliable, although few topics are as contentious in modern evolutionary biology as the evolution of sex. Some mathematical models show that sexually reproducing organisms evolve faster, and some show that they evolve slower. Some show a clear advantage to sex when the environment is rapidly changing, because should an ecological disaster occur,

---

* *The Red Queen: Sex and the Evolution of Human Nature* by Matt Ridley is a comprehensive treatment of sex and evolution in humans.

the shuffling of inherited information means that someone, somewhere is likely to have the solution. Aphids (the tiny green insects that often infest garden plants) reproduce asexually during the summer, efficiently churning out clones that are identical to themselves. In the autumn, however, they reproduce sexually, laying eggs with more variation of characteristics than either of the parents. That way, presumably, an unexpected winter environment will not kill the entire clutch.

No explanation for the evolution of sex is universally accepted. But whatever the reason, most scientists would agree that sexual reproduction was essential for the diversity of life on Earth. Even more than that, it seems unlikely that life on Earth would have reached anything like the complexity that it has, were it not for sexual reproduction. No animals, no plants, possibly not even any amoeba, which, it turns out, aren't as asexual as scientists used to think. Even asexual bacteria exchange genes with each other, handing over useful genetic information like secret agents dropping off coded messages on park benches. So while we can't say anything about whether alien planets have sexually reproducing organisms or not, we can say that if they have complex life, it likely arose through some 'accelerated' natural selection, and possibly a process similar to what we on Earth call sex.

This kind of reverse reasoning about evolution must be approached with caution; we cannot say that we expect alien planets to have sexual reproduction because it's 'good' for evolution. Rather, we can say that evolution might not produce complex animals without some similar process. For decades astronomers (among them Carl Sagan) have speculated that mysterious dark patches in the atmosphere of Venus might be airborne colonies of microorganisms. Maybe. If at least two planets in our solar system have life, then other planets around other stars may well be swarming with simple life. But only simple life that never evolved into anything more complex than bacteria.

The second point that is important to us is that sex is hugely significant – perhaps unexpectedly significant – in the evolution of complex *behaviour*. Asexually reproducing organisms such as bacteria have very little in the way of interesting behaviour, and not much in the way of social behaviour in particular. Of course, that may just be because they are physically simple (or that we haven't yet uncovered their secret social lives, although see Chapter Seven for some details on bacterial cooperation), but there is another mechanism at play. In a society composed of genetically identical clones, as the offspring of asexual reproduction must be, everyone is equally related to everyone else, and so there is no opportunity for kin selection to act and favour more complex behaviours.

When sexual reproduction takes place, relatedness conflict arises. I am more related to my children than to my nieces and (with apologies to my sister) I am more likely to support my own children than hers. Of course, human societies and relationships are more complex than this, but nonetheless, this asymmetry leads directly to family units, greater social structure and the whole panoply of complex behaviour that is almost ubiquitous among animals. If my father cloned me and my sister to be genetically identical to him, and we cloned our own children, I would be just as related to my two nieces as I am to my children, and in fact just as related to them as I am to myself! In such a society, everyone would be happy to help anyone else, without structure or differentiation. Ironically, it is the complex network of conflicts of interest that generates different roles for individuals, and the emergence of what we regard as sociality. This peculiar fact is the subject of Chapter Seven.

So the evolution of life with anything like the vast diversity we see on Earth probably requires some kind of sex. But what is sex? Can we generalize the concept so that we are free of any of the peculiar evolutionary history of Earth, and in particular free of specific concepts like DNA, which are merely a coincidence of our biochemistry?

The most important feature of sex, from an evolutionary perspective, is that your offspring get some of their inherited characteristics from someone else. That doesn't imply that there are just two sexes, and it doesn't imply that there are only two parents. A lot of fungi have many, possibly thousands of different sexes, which is obviously advantageous if you want to be sure you can mate with anyone you meet (they're unlikely to be of the same sex as you). Multiple parenting is not common on Earth, but the recently well-publicized case of a child born to three parents shows how our ideas of simple inheritance by sexual reproduction aren't quite as inevitable as we might think. In the case of a child with three parents, the insight is that we actually have two different sets of DNA in our cells, one combined from our parents, and one from just our mother. Therefore, an alien organism with multiple sets of inherited information from multiple parents is not that difficult to imagine. So it seems that sex as it appears on Earth is no more than a particular form of a very general function: the shuffling of inherited traits between individuals. The function, which seems to speed up evolution, avoids parasitism and ensures against ecological disasters, could be implemented in different ways on different worlds. But if sex, or something similar, takes place on alien planets, the process of evolution is surely very similar in both places.

We have laid out our radical approach. We *can* talk about what aliens are like, because the rules of evolution are similar on all planets. More than that, we observe on Earth the ubiquity of convergent evolution, even between the most distantly related organisms, and conclude that these organisms would also converge in their functions if they lived on different planets.

The principle of convergence is so powerful, and so simple, that it is hard to argue that similar challenges will not produce similar solutions. Similar solutions will arise multiple times because *we live in a universe where not everything is possible.*★ If inhabited planets are radically different from Earth in their physical and chemical nature, possibly much hotter or much colder too, then we can't expect similar forms to those we see on Earth – feathers are for flying through air, not through the ammonia clouds of Jupiter – but we should not be surprised to find the same functions (i.e. flight) that we observe here.

Of course, a nagging worry is that this convergence is somehow a side effect of the biochemical similarity of all life on Earth. We share even with bacteria the genetic code based on DNA. But it is stretching credulity to breaking point to suggest that it is the very nature of DNA that is responsible for convergence. The fundamental forces at work in evolution are independent of the precise details of which molecules interact with which and how. I have said very little about biochemistry, and that might surprise some readers, as most books on astrobiology deal, for the most part, with the different kinds of molecules that could make up life and how those molecules came to exist in the first place. Must life be based on DNA? Must it even be based on carbon chemistry? Is liquid water essential for life?

These are fascinating questions with complex answers, and there are many books detailing the current state of research.†
But stepping back from the vital question of how life *exists*, we

---

★ *Life's Solution* by Simon Conway Morris.

† More technical books include *Astrobiology: Understanding Life in the Universe*, by Charles S. Cockell, but there are also many more popular books such as *Life in Space: Astrobiology for Everyone* by Lucas John Mix.

can see that evolution tells us about how life *develops*. Remarkably, the answers to these questions are very similar whether your metabolism works on liquid water or liquid methane.

So when we look at animals on Earth in all their diversity, particularly their diversity in function, if we squint then it isn't hard to imagine, in their place, a huge variety of aliens. When Mary and Joseph Anning uncovered their fossil ichthyosaur, they could infer much about its lifestyle and behaviour, despite the fact that no human had ever seen such a creature alive. In almost every way, this was an 'alien' animal. Nonetheless, the ichthyosaur, and its parallels on other planets, achieved their particular characteristics through the relentless and often predictable action of natural selection. Looking past the outward form of animals on this planet, and focussing on how they live and interact, they are probably not really so different from their parallels across the universe. The rest of this book is about how we can make quite specific predictions about what those parallels will actually be.

3.   What are Animals and What are Aliens?

I talk to spiders in my bath. I plead with them to crawl to safety, but they don't listen to me. To me, it seems obvious that I'm trying to help them when they are stuck in a creepy white valley, surrounded by slippery precipitous cliffs from which they can't possibly escape; but they certainly don't realize that I'm on their side. If I offer them a piece of paper, they won't climb onto it, or if they do, they jump off again as soon as I begin to hoist them out of the bath. All the time, I keep trying to convince them to let me help. Why? Many years ago, I read that the best way to get spiders out of a bath without harming them is to drape a towel over the edge, so that they can climb up and out. I rarely do this – not because I'm worried they'll end up lurking in the towel, but because it seems impersonal, indifferent. I *want* to talk to them, to explain my good intentions.

My behaviour is odd but not unprecedented. We have an urge to talk to animals, even if they can't understand us. For us, the status of 'animal' holds some special significance. But what do we think an animal *is*? We seem to recognize a bond between us and all other animals. But how do we know what makes an animal . . . an animal? Why are we prepared to talk to a spider, but not to a mushroom? Indeed, we seem to be prepared to talk to, and assume a level of understanding by, creatures that are as radically different from us as we can imagine. Even an earthworm is a worthy partner in conversation (albeit a rather one-sided conversation). Would we feel the same way about aliens? In fact, is it likely that alien life forms would be recognizable as 'animals' at all?

We have an instinctive understanding that we share something in common with all animals – we're not sure what that is, and it certainly isn't as simple as physical appearance, but we know that you and I, and the spider and the earthworm, all are part of the same group that we call 'animals'. Can we – the general public as well as the scientific community – agree on what animals are? And even if we can agree on what animals are on this planet, would that definition apply to animals on other worlds? Agreeing on what defines an animal is something with which humanity will have to come to terms sooner or later. We have a bond of identity with animals – whatever they are – and that affects our attitudes towards them, both ethical and social. And in case you are surprised that 'what is an animal?' is still a question, consider that the Kentucky State Legislature still defines 'animal' as 'every warm-blooded living creature except a human being'.* That is, reptiles and fish are not animals in Kentucky! As a result, they are not afforded the legal protection against cruelty that is extended to mammals.

Defining 'animal', by form or function?

Defining what is and is not an animal is a very ancient problem. Aristotle himself struggled with the question of how to classify a sponge (he decided – correctly, as it happens – that it is an animal). With time, we have refined how we view the divisions between the different types of life, and as our technology improves (microscopes, chemical analysis, DNA sequencing), we believe that we are closer to a kind of 'true' classification. But what has worked spectacularly for natural scientists on Earth, may not bring us the results we need when we eventually have to apply these principles to life on other planets.

* Kentucky Revised Statutes [KRS 446.010(2)].

Over the centuries, there have been two ways that scientists have tried to classify life. Beginning in ancient Greece with Aristotle, and continuing with diligent and precise observers of nature during the Renaissance and Enlightenment, people have tried to classify organisms by their *form*: what they look like, how they live, what they're made of and so on. This straightforward approach – the one that we are often taught in school – has appeal, because we use a similar structural/functional approach to classify most inanimate objects in the world: if it opens a bottle of beer, it's a bottle opener. Recognizing similarities between objects is how we learn to generalize as young children: that is a dog, and this is also a dog, even though they're not the same dog. In this way, scientists from ancient times onwards have tried to come up with clear-cut rules that divide organisms into unequivocal groups: if it's a dog, it's not a cat.

Clearly, this could be a minefield for astrobiologists. Aliens are likely to be very different from animals on Earth, but they might also have many similarities. Just because a young child points at an alien and says 'Dog!', that does not necessarily mean that we should extend our definition of dogs to include these alien canines. Something more than form alone is needed to provide robust classification. The diverse range of creatures we see must be considered in its entirety, so that we can also take into account the relationships between groups, as well as between individuals.

Careful study of the similarities between animals and plants, as well as their differences, led the Swedish biologist Carl Linnaeus to propose an elegant system based on *hierarchical* classes: a dog is a carnivore, like bears and tigers and raccoons, and like all of those a dog is also a mammal, and mammals are all animals, and so on. So powerful was this system of assigning hierarchical groupings to all life, based on similarities in their form, that Linnaeus's basic structure has remained in place to this day. Nonetheless, this structural/functional approach to classifying life raises interesting dilemmas. For example, for almost 2,000 years, scientists followed Aristotle in defining animals as having five crucial characteristics: movement, sensation, digestion, reproduction and physiology (i.e. some kind of internal mechanism that makes them 'work'). However, many creatures that very much appeared to be animals, such as oysters, appeared to be lacking one or more of these characteristics. In particular, reproduction in marine animals is not always that easy to observe, and so people doubted whether shellfish were in fact animals at all, because they could not see them mating or giving birth.

This is a warning message for the study of alien life: we should not assume that reproduction, for example, takes the familiar form we see in mammals and birds, with males impregnating females, who then carry the developing offspring, at least for part of their developmental cycle. If we haven't considered that fertilization can take place outside of both bodies, we may not recognize it as reproduction at all.

Comparing species by what they do or don't have in common is also complicated when we can't be sure which features are the most important, particularly when they lead to conflicting conclusions. Birds fly, and bats fly: are bats therefore birds? As you'll remember from Chapter Two, we don't think so: they don't really look like birds, because they have fur instead of feathers, and they give birth to live young instead of laying eggs. Here the problem is more fundamental than a mere inability to find a good set of rules that divide up animals in a way we find satisfying. How can we be *sure* that bats are not really birds? Which is more important for determining a category: having wings or having fur?

A similar and even more contentious debate raged for hundreds of years over how to classify dolphins and whales. It is easy to laugh today at Aristotle's claim that whales are fish: they patently breathe air and give birth to live young just like other mammals. Nonetheless, Aristotle was a rigorous and systematic observer, and he carefully recorded all of the facts about whales and dolphins, including their air-breathing habit, with lungs and blowholes, their warm blood and the way that they feed their young with milk-providing mammary glands. He remarked on how similar dolphins were to humans and horses, but lacking the evolutionary perspective that we have in hindsight, he felt that having no legs was just too big an obstacle for our aquatic relatives to be placed with the remainder of the mammalian world.★ Thus, 'fish' seemed a more natural category for the dolphin and the whale. Scholarly observers of nature concurred with this assessment, right up until Linnaeus gave a holistic reassessment of the similarities and differences between mammals and fish, and thus in 1758 decreed whales and dolphins to be mammals. The appeal of structural/functional classification is strong, but precise boundaries can turn on the relative importance of small differences. Is it more important that whales breathe air (making them mammals) or live in the sea (making them fish)? One hundred years later, the discussion was still not closed; Herman Melville wrote in *Moby-Dick* (1851):

> In some quarters it still remains a moot point whether a whale be a fish . . . Be it known that, waiving all argument, I take the good old fashioned ground that the whale is a fish, and call upon holy Jonah to back me. This fundamental thing settled, the next point is, in what internal respect does the whale differ from other fish. Above, Linnæus has given you those items. But in brief, they are these: lungs and warm blood; whereas, all other fish are lungless and cold blooded.

Like many people today, Melville mistakenly assumed that Jonah's nemesis was a whale, although the Bible only refers to it

★ See *Animal, Vegetable, Mineral? How Eighteenth-Century Science Disrupted the Natural Order* by Susannah Gibson.

as a 'big fish'.★ It seems unlikely that when the book of Jonah was written 2,500 years ago, people either knew of or cared about the difference – it looks, after all, like a big fish! However, Melville's important point is that whales really can be considered just fish with lungs and warm blood. There is no logical inconsistency here, and why not consider whales to be fish – if we choose to accept classification on the basis of appearance. It all depends on which features you consider more important than others; in this case, physiology or behaviour.

And so it seems that organizing animals according to their form – appearance or structure – can't always deliver the clear boundaries that it promises. There simply seem to be too many exceptions to any rules we think of. 'Birds fly' – except ostriches, penguins, emus, kakapos and various others. 'Mammals don't lay eggs' – except platypuses and echidnas. In fact, the reason why there are so many exceptions to rules of appearance is only to be expected, given the way that evolution progresses. Birds fly, but some of them have latterly evolved to shun flight, because it serves them better to swim or run. Mammals don't generally lay eggs, but they evolved from animals that did, and so some of their descendants still do today. What makes birds 'birds', or mammals 'mammals', is their evolutionary heritage: they descended from bird ancestors, or mammal ancestors. But it wasn't until the 1800s that someone came along who could make that huge leap of insight.

The new definition: pedigree

Just 150 years ago, and almost overnight, our method and even our rationale for classifying life was turned on its head. Once

★ Jonah 1:17, 'The LORD provided a huge fish to swallow Jonah; and Jonah remained in the fish's belly three days and three nights.'

Darwin declared that all modern life is descended from earlier forms of life, the classification of living organisms (or 'taxonomy') developed a new emphasis: classification should be based on heritage, not appearance, form or function. The question is not 'Does a whale behave like a fish?' Rather, we need to ask, 'Is a whale closely related to a fish?' Although this now seems to us an obvious approach, the very principle that a whale might be related to a frog, or even a mushroom, however distantly, would have been ridiculous to Aristotle. But consider the elegance of the new method. The old definitions of the 'Kingdoms' of life – Animals, Plants and Protists (microbes) – were retained, but took on more than just descriptive meaning. They now represented the roots of separate family trees; the patriarchs and matriarchs of a genealogical system that was easily understandable in terms of human ancestry. Like in Shakespeare's *Romeo and Juliet*, if you belong to the Montague family, you're a Montague, together with all your Montague relatives – Capulets have their own separate family tree. But at the same time, all these family trees are linked together into a larger Tree of Life. Even the warring Montagues and Capulets must admit that at some point they shared a common ancestor.

Extending this principle, the fact that we share a single common ancestor with all life on Earth means that hard distinctions between any two groups of organisms no longer exist. We can even calculate how far back in time we have to go until we share a common great-grand-ancestor with any other species. The family tree below gives you some idea of just how far back this ancestor is for some well-known creatures, such as a wolf. In this case the shared ancestor was a small furry mammal (about which we know very little) that lived at the same time as *Tyrannosaurus*, and of its offspring, some evolved to be carnivores, and others to be primates like us. To find a common ancestor with a goldfish, we have to go much further back – long before the time of the dinosaurs. More distant still, our shared history with plants and fungi is lost in a time before the earliest of fossils. Even insects diverged from our own ancestors just as the very first animals were finding their own niches and diversifying into myriad forms.

We can estimate this family tree only through careful study of the similarities and differences in the genetics of the different species today, but there remains much uncertainty about the dates themselves, and more scientific studies will help us focus in on exactly what happened in our ancestry and when. Our shared heritage with bacteria like *E. coli* goes back to the dawn of life itself, not long after the Earth cooled enough for the first complex chemicals to form. Right at the root of the tree presumably sits our Last Universal Common Ancestor, or LUCA (no relation to 'Luka', the 1987 song by Suzanne Vega, although the lyrics are particularly pertinent: 'Just don't ask me what it was').

Whether there was actually a single common ancestor, or something more complex (multiple bacteria-like organisms sharing their DNA in a kind of genetic commune, perhaps), the use of the word 'universal' here is definitely misleading. Life's ancestor was quite likely an Earthling, and so is unlikely to tell us very much about the common ancestors of life in other parts of the galaxy. Even if life originated on Mars and was transported to early Earth by meteorites, as some scientists have proposed, it is still very much life from our own solar system, with no relationship to the billions of other planets in the galaxy.

But uncertainty as to the precise roots of the tree does not detract from its appeal as an elegant and, frankly, *valid* way to classify life. So appealing was the family tree approach that, once accepted, it has never been challenged. But challenged it must be. Because, barring any bizarre transfer of living material from one planet to another, we can be sure that life on other planets does not share a common ancestor with us, or any other creature on Earth. A classification based solely on heritage, and not on

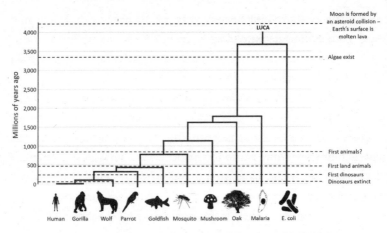

A family tree showing how long ago we shared a common ancestor with other familiar species. At the very top, 4 billion years ago, is the ancestor of all life, LUCA. Each branching shows when different groups of organisms went their separate evolutionary ways. These numbers are taken from the website http://www.timetree.org/ but are always being updated with new research. Note that the uncertainty on some of these dates is very large – and larger for the earlier branches (as much as ±2.5 billion years for *E. coli*).

form or function, would exclude all aliens, no matter how similar to us, from the category of 'animals'.

Does that matter? Probably. Innumerable science fiction stories portray alien species as animal-like: the sandworms of *Dune*, Edgar the Bug from *Men in Black*, and just about all the aliens of *Star Wars*, especially Chewbacca and the Ewoks. However realistic, or unrealistic that may be, our reaction to animal-like aliens is going to be emotionally similar to our reaction to animals, irrespective of whether or not we share a common ancestor. The technical question of should we classify an alien as an animal is nothing compared to the ethical and emotional question: should we *treat* an alien as an animal? It may be more logical, and more useful for the purposes of astrobiology, to blend consideration of both the evolutionary heritage of organisms and of their properties, when considering how to classify them.

Is family everything?

We already recognize that relatedness isn't the only criterion by which to judge our relationship with other animals on the planet. The last ten years have seen a flurry of lawsuits attempting to assign 'human rights' to non-humans, most famously when a court in New York was asked to grant a writ of habeas corpus to two chimpanzees, Hercules and Leo, being used for research at Stony Brook University.[*] Although it seems that society (and the courts) are not yet ready to grant full human rights to animals (a topic I discuss more in Chapter Eleven), laws protecting animals as animals (i.e. to protect them against abuse and cruelty) have continued to strengthen. But now we are faced with a new dilemma: how do we decide which animals are worthy of protection, and which can be exterminated at will? The Stony Brook court case is an illustration of how strongly we feel about our closest relatives – most people would feel uncomfortable watching a chimpanzee suffer, because they are both closely related to us (sharing a common ancestor just 6 million years ago), and similar to us in their behaviour, expressions and, most likely, their emotions.

Consider dolphins, however, which attract universal empathy for their superficially human-like behaviour. In 2013, the Indian government decided it would no longer allow dolphins and whales to be kept in captivity for entertainment, noting that 'various scientists' have suggested that dolphins should be considered as 'non-human persons'. Yet dolphins are less closely

[*]https://www.nonhumanrights.org/blog/hercules-leo-project-chimps-sanctuary/.

related to us than rats, and family ties with humans do not help rats escape routine extermination. Dolphins, for all the behavioural properties they share with humans, are more distant relatives of ours than, for example, vampire bats, who also get very little good press. In fact, vampire bats are very sociable and altruistic animals, who share their food with hungry individuals who haven't managed to feed (see Chapter Seven). Looking deeper than just outward behaviour, we can also see that dolphin brains are in many ways more similar to the brains of bats than to the brains of humans. This may come as a surprise to those who assume that dolphin intelligence must be reflected in human-like brains, but of course it is an important wake-up reminder when we come to consider whether or not alien creatures 'like us' will also have brains like ours.

Dolphin brains and bat brains are similar because both species need incredibly complex neural apparatus to process their own unique forms of sonar. They produce complex sounds and then process the echoes they receive to find out about the world around them, without the benefit of good eyesight. This processing of sensory information takes place in a part of the brain called the cerebellum, which is proportionally larger in bats and dolphins than the much smaller cerebral cortex – the part of the brain usually associated with intelligence. This emphasizes that physical similarity – especially the physical similarity of brains – does not really give a good indication of how similar to us a species might be.

In 2010, the European Union enacted a strict set of legislation restricting the use of animals for laboratory experiments. The directive applies to all non-human vertebrates – mammals, birds, reptiles, fish and amphibians – but not invertebrates like insects and worms, which are not considered cognitively developed enough to 'suffer' in a way that humans understand. However, the law makes special exception to include octopuses and their relatives because of the way they clearly demonstrate intelligence and, most likely, sentience. Octopuses have been regularly reported to remove their aquarium lid, cross the floor to another tank, eat the fish inside, and then return to their own tank, replacing the lid as they go. Although the EU directive is based on the suspected ability of octopuses to feel not just pain, but also boredom, nausea and fear, it would seem that we personally identify with the consciousness of these creatures. The octopus and its relatives (collectively known as cephalopods) are the subject of our empathy and concern, despite the fact that our last common ancestor with the octopus was 800 million years ago. We are about as distantly related to the octopus as we are to any other animal! When it comes to defining animals as being 'like us', familial closeness is clearly not the only factor. So, is there a definition of animals that does not rely on our family tree?

Animals as animated

Science needs to classify the world around us, and so science has come up with simple and unequivocal criteria for what all animals have in common – these are the criteria you will learn at school or university. Animals are first and foremost made up of many different cells, and in almost all animals, different cells perform different functions. We have skin cells and blood cells – they look different and they behave differently. That definition alone is enough to distinguish animals from microscopic life forms like bacteria and amoebae. But of course, plants have lots of different cell types too. However, animals are different from plants in one vital respect: animals don't make their own food. Whereas plants can use sunlight to turn simple molecules like water and carbon dioxide into the chemicals they need to survive, animals have to go out looking for food, and it is this 'going out' that really makes animals different. Fungi, a hugely diverse group of organisms (there are about twice as many species of fungi as there are of plants) also can't make their own food. But unlike animals that ingest their food and process it inside their bodies, fungi digest their food externally by secreting chemicals that break down other life forms into the nutrients that the fungi need. The only time fungi move is as spores, dispersed into the wind; once they settle, they consume whatever they find wherever they land.

This is a really good distinction between animals and every other form of life on Earth, because it fits in well with our intuitive understanding: animals *move*. Although the word 'animal' comes from the Latin word *animus*, which referred simply to something being alive or breathing (think of 're-animation'), we tend to think of something 'animated' as being something which moves, acts and takes control of its situation. So, although the scientific answer may be that animals are multicellular organisms that cannot make their own nutrients but ingest material that is broken down inside their bodies, we tend to short-circuit all that and, like Aristotle, simply say that 'animals move'.

The real world is full of exceptions: some animal species only move in their reproductive form, such as coral larvae that are dispersed into the water and drift around until settling, where they remain for the rest of their lives. In that sense, their 'motion' is rather similar to that of dispersing fungal spores, or plant seeds. But we shouldn't be afraid of exceptions; the classification of life is never going to be simple and clear-cut, because life itself is almost infinitely gradated, and binary distinctions are in fact the exceptions themselves, rather than the rule. Instead, I want to focus on a more difficult question: can we be confident that our definition of animals truly represents a distinct category of organisms, and isn't just a coincidence, reflecting the very particular and specific history that life has lived through in the last 4 billion years on Earth? In short, will animals on other planets move?

Can it be as simple as that? Animals move, plants do not. It seems intuitively naïve, a trivial distinction that may apply on Earth only; something not generally applicable throughout the universe. Perhaps I'm using the term 'animal' too loosely; perhaps I mean 'something with which we can identify', like the spider in my bath. What I would describe as an 'alien animal' might have leaves and might photosynthesize, rather like the character Groot from the *Guardians of the Galaxy* franchise. But I believe – and as I will show – the ability to move underlies all the other developments I will discuss: cooperation, sociality and – especially – intelligence.

There are good reasons to think that the distinction between movement and no movement is more fundamental than mere whim. But to answer this question, we have to go back to the time when animals first evolved, between 500 and 1,000 million years ago, back to when we share our common ancestry with octopuses and mosquitoes.

The Garden of Ediacara

Until the 1950s, palaeontologists generally thought that complex life began around 540 million years ago in what was called the Cambrian explosion. There are almost no fossils from before that time, but an 'explosion' of variety in complex fossils from then onwards, showing diverse forms of animal life, many of them sharing fundamental features with life today: symmetrical bodies, appendages, eyes and so on. Almost all the basic categories of animals that are alive today (molluscs, crustaceans, worms, even primitive vertebrates) exist in fossils from between 540 and

490 million years ago. So what kind of life existed *before* the Cambrian explosion?

The discovery of delicately preserved, sometimes microscopic fossils from before the Cambrian has given us a crucial, but still hazy, window through which to understand the origin of animal life. This time between 630 and 540 million years ago is known as the Ediacaran period (named after the Australian hills where such ancient fossils were first discovered), and some of the hypotheses about the nature of life during that time are quite revolutionary. Ediacaran animals did not seem to have shells of any sort – that is one reason why their fossils are so rare – and their structure seems very different from almost any other animals alive from the Cambrian until today. How this peculiar and alien-looking set of life forms was suddenly supplanted 540 million years ago, leaving no descendants for us to study, is a mystery.

One explanation is that this ancient world was almost completely without predation. These creatures lived when nature was not 'red in tooth and claw', as Tennyson observed; they lived in a time without conflict, peaceful and serene, when creatures simply collected energy from the sun. Possibly these creatures were themselves photosynthesizing, or possibly they had symbiotic relationships with microscopic algae living inside them, as some modern corals do; animals harvesting the sunlight with the help of non-animals. In any case, their foraging tactics appeared to be quite harmonious, rather than hunting down and ingesting each other. This vision of a peaceful world without predators led the American palaeontologist Mark McMenamin to dub this era 'The Garden of Ediacara'.*

Let us then walk through the history of life up until this

---

* *The Garden of Ediacara: Discovering the First Complex Life* by Mark A. S. McMenamin.

point.* Somehow, at some point, life began. We don't know what that event was, how it happened, or even whether it happened in one place and spread, or happened in many places at the same time. But most likely, small replicating molecules, similar to modern RNA, began to spread through the warm ponds that covered the Earth. Soon, genetic material became encased in a fatty bubble, which protected it from changes in the environment, which we might call the first cells. There may have been many different types of primitive cells, or perhaps just one, but one organism (or possibly one group of organisms) outcompeted the others, and became the sole ancestor of all subsequent life on Earth: our Last Universal Common Ancestor, LUCA.

As LUCA's descendants spread and diversified, they all faced similar challenges: primarily, where to get energy. Without energy, life decays and ceases to be life. The heat of the Earth, coming up from volcanic activity under the sea, was one source, and the light from the sun was another. There really weren't any other alternatives at that time. Sunlight, however, was available all over the Earth, whereas underwater volcanoes only occurred in certain places. Therefore, these organisms evolved chemicals and organs to capture sunlight and use that energy to keep them alive. When you are lucky enough to find a remote seaside beach, there is plenty of sunlight and no competition for places to sunbathe, so it is a relaxing, stress-free holiday. However, a lack of competition does not bode well for evolution – as in those circumstances there is no advantage to innovation, and there are no real problems that need to be solved. Some organisms fared

* Richard Fortey's book *Life: An Unauthorised Biography* provides a diverse and very readable account of the history of life on Earth, although it doesn't deal specifically with the origin of life. *The Origins of Life: From the Birth of Life to the Origins of Language* by John Maynard Smith and Eörs Szathmáry is far more detailed, but much more technical.

An artist's impression of life in the Garden of Ediacara. Some of these creatures may be animals, or they may not. But judging by the lack of protective spines or shells, or legs with which to run away, a kind of relaxed complacency seemed to be the order of the day.

better than others and reproduced more, but everyone had a pretty easy time and life forms were pretty simple. Life began about 3,800 million years ago, but for the first 3,200 million years, no one ate anything other than sunlight. Evolution was slow. Nonetheless, evolution was still taking place, and the Garden of Ediacara contained a huge range of different life forms, each, presumably, exploiting the peaceful and plentiful environment in slightly different ways.

It's not clear whether the Ediacaran fossils are animals or not. That uncertainty centres on the family tree definition of what an animal is. At some point, an organism existed that was the common ancestor to all modern animals (and also, bizarrely, happened to be the ancestor to fungi). This creature was single-celled and propelled itself with a tail, rather like a sperm cell. It bears the fantastic name of opisthokont – well worthy of a creature with such a foundational role as the ancestor of all animals. Opisthokonts appeared around 1,300 million years ago, long, long before Ediacaran times. Was it also the ancestor of the creatures whose fossils we see in the Ediacaran rocks? If so, then by the family tree definition, that would make them animals. If opisthokonts were a separate branch, then some further back ancestor was common to both animals and Ediacarans.

Whatever their family history, though, these Ediacaran creatures did not seem, by their properties, to be very animal-like. They sat around soaking up the sun and, although it seems they could move, they do not appear to have moved with much urgency. Fossils from after the Garden show lots of evidence of animals digging into the sand to hide, or scurrying over the seabed, but such signs of frantic activity are totally lacking from Ediacaran times. It was chill – so chill that these ancestors of animals lacked the *animation* that we inevitably associate with the definition.

There's no guarantee that this course of events bears any similarity to the course of early life on another planet. And yet, there's nothing in this story so far that seems particularly specific to Earth. Sunlight (or light from the equivalent star on another planet) is quite likely to be the most available, most powerful source of energy for sustaining life. Different ways of exploiting that energy may evolve. At first, at least, many planets will go through a similar phase of diverse life, based solely on free starlight.

Leaving the Garden of Eden

Suddenly, everything changed. Maybe the beach filled up with sunbathers. More likely, the climate changed and the endless free lunch was no more. One creature began to get energy, not from the sun or from the hot volcanic waters, but from ingesting other living organisms. In the biblical story Adam and Eve, and all the other creatures in the Garden of Eden, were vegetarians, living the easy life, until they were thrown out of the Garden; and so it was in reality.* Once predation became a thing, evolution went into overdrive: those organisms that didn't adapt were lunch – literally. All kinds of defensive and offensive traits began to appear: protective spines and shells, teeth to take a bite out of someone else, eyes to find who to bite – or spot who is going to bite you. Within a few tens of millions of years, the Garden of Ediacara was no more. Nature was all tooth and claw, if not yet actually red with blood. These, then, were the real animals, running, swimming and burrowing for their lives, in contrast to the plants of the time, who stubbornly sat still and continued to absorb sunlight, dealing with the relentless assault of grazing creatures as best they could.

This tale of life on Earth is, in some ways, very specific to our own planet, but in some ways, it is very general indeed. True, you would not expect to see the precise same sequence of events and timings on another planet as the events that led to the Cambrian explosion on Earth. Nevertheless, some features are the same. Evolution requires pressure, competition, scarcity. Idyllic gardens are unlikely to lead to life forms with the complexity we see today. We can be sure that, throughout the universe, life will require at least two things: energy and space. Energy, because the laws of physics state that without a constant input of energy, systems decay and become disordered – the opposite of life. Space, because two organisms take up more space than one, and reproduction is central to natural selection – the only mechanism we know of for

---

* Genesis 1:29: 'God said, "See, I give you every seed-bearing plant that is upon all the earth, and every tree that has seed-bearing fruit; they shall be yours for food."'

complexity to arise on its own. Eventually, there will be competition for energy and space. Animals – creatures that move to compete for those scarce resources – are virtually inevitable.

Universal animals

The problems facing animals on Earth are universal problems. Eat. Avoid being eaten. Find space to live. Reproduce. Did life on alien planets find alternative solutions to those that we see on Earth? Perhaps it is a mistake to draw too many general assumptions and to conclude that animals are inevitable throughout the universe. But I think not. One of my undergraduate professors, the Cambridge palaeontologist Simon Conway Morris, was one of the first people to describe the very earliest animal fossils in great detail. He argues strongly that evolution often seems to come up with similar solutions to the same problem.* Wings of bats and birds are utterly different in their structure, but they are wings nonetheless. The eyes of vertebrates are very different from insect eyes, but in contrast, the equally distantly related cephalopods evolved a very similar visual structure to us, and quite independently.

Simon Conway Morris speculated on an alien planet inhabited by intelligent insect-like creatures with compound eyes that form very low-resolution images. As their technology improved and they developed astronomy, eventually building telescopes according to the same physical principles as on Earth, surely they would understand that a single lens forms a more precise image than their own compound eyes? Being good scientists, they would be humble enough to predict that on another planet (Earth, say), animals exist with lensed eyes, rather than compound ones like theirs.

* See *Life's Solution* by Simon Conway Morris.

Even though these alien astronomers would be different from us in so many ways, they would understand that the laws of physics constrain their abilities just the same way as ours, and those constraints are true on any planet. If alien compound-eyed scientists can speculate about lens-eyed Earthlings, why shouldn't we be able to speculate about alien animals vastly different from us?

Perhaps the range of evolutionary solutions we see on our planet is closer than we think to the range of solutions across the universe. In our justified care to ensure we do not extend analogy too far from our Earth-bound experience, we must similarly be careful not to over-enthuse on the most unlikely alien solutions to the same problems. If we see, as appears to be the case on Earth, that not all of the possibilities are exploited, it may be because some solutions are simply not available; perhaps they are impractical, inefficient or physically impossible: famously, for example, no animals have evolved wheels. As Conway Morris says, 'We live in a constrained world, where all may not be possible.'

Given, then, the likely scenario that once competition (and hence, evolution) hots up, movement may be the only practical solution to gathering resources, we expect to find moving creatures on any planet where complex life exists. Nonetheless, other inhabited planets may be unimaginably different from our own – seas of liquid methane (like on Jupiter's moon Titan), or diamonds raining from the sky (as may be the case on Uranus and Neptune). Can we be sure that whatever the conditions, we would recognize the *moving* organisms of these worlds as animals? This seems very likely, because whatever the physical difference between planets, the fundamental processes are the same. Natural selection is our shared heritage with life throughout the universe. Specific solutions to the problem of, say, how to move certainly depend on planetary conditions.

If evolution does indeed push life on all planets towards organisms that move around to compete with each other, is this a helpful answer to our question, 'What is an animal?' Partially. There can be no simple and unequivocal resolution to the dilemma over whether we should classify creatures according to their outward appearance, or to their heritage. Some humility is in order when examining our capabilities to answer this question. However, if you want to know why I talk to the spiders in my bath, and why I want to talk to alien animals, it is not because of our shared ancestry, but because of our shared properties – the things that make us all animals.

Although it is quite possible that one day we will discover an alien planet still bathing in the glory of its own Garden of Ediacara, the basis of the moving, struggling, bustling future will already have been laid. And the laws governing that movement are the subject of the next chapter.

4. Movement – Scuttling and Gliding Across Space

We have an instinctive fear of scuttling creatures. So it's no surprise that science fiction aliens often pursue their victims on multiple legs, or with a quirky jointed movement. We fear the chase of predators, and with good cause. To survive, we must move, and so must both our victims and our enemies. But beyond the mere essence of movement, what can we say about the different kinds of animal movement that might exist on other planets? On the face of it, understanding how alien animals might move is a relatively simple quest. Movement is essentially a problem of physics, and the laws of physics are universal: forces and acceleration, torque, friction and so on, are the reality on every planet in any solar system. But we remain uneasy about such grand declarations. Perhaps there is a novel way to move around that we haven't thought of. Perhaps some planetary environments have physical properties so unique that we haven't even considered how animals might move around in those conditions.

That is why we must start by considering not the physical constraints on movement, but the evolutionary ones. Movement exists in the animal world only because of evolutionary pressures. We move because we must, not because we can. Physical constraints certainly influence how we can move, and even sometimes if we can, or should, move at all. Plants survive without moving around (mostly). But for those animals that move, the movement itself arises out of necessity. Animals on planet Earth show such a huge diversity of movement mechanisms that we can be tricked into thinking that there must be equally many different reasons to move, but actually this is not the case.

Why do animals move?

Animals move to find food, of course, and to avoid becoming someone else's food. But more generally, we can identify the drivers of movement as the three resources that are universally limited: energy, space and time.

The first life forms on Earth for which we have any fossil record are usually thought to be stationary. These ancient fossils from at least 3 billion years ago are so similar to modern structures called stromatolites that the general assumption is that they were produced by similar creatures, very simple bacteria that take their energy from the sun and grow in 'mats' that emerge from the water as new bacteria grow on top of previous layers. These bacteria don't 'move' in the animal sense, but the colony as a whole grows upwards, because as sand and dead organic matter accumulate, their source of energy – the sun – would be blocked out. So even for ancient bacteria, movement in search of energy was essential, just as trees grow upwards to avoid being shaded out by their competitors.

Alien worlds might possess sources of energy unfamiliar here on Earth, or at least less common. Underground oceans like those on Saturn's moon Enceladus receive no sunlight, but there is energy aplenty, partly from the radioactive decay of elements in the planet's core, but also from the friction of the tides – the huge force of the gravity from Saturn pulling the rock and water backwards and forwards. Life may well be present on these worlds too, and if so, it will need a way to find and harness these unfamiliar energy sources.

Life needs energy, and if energy is not evenly distributed, life

Modern stromatolites growing in Australia (left), and a cross-section of a fossilized stromatolite, showing the layers of bacterial mats (right).

must go in search of it. Of course, sunlight is by and large present everywhere on Earth, and so photosynthesizing organisms don't really need to move anywhere but upwards to get energy. However, when all life is competing for a single energy source (the Sun), then alternative strategies begin to look attractive to evolution. An organism that waits for bacteria to gather solar energy, and then gobbles up the bacteria, doesn't need to compete for sunlight; something like this appears to have happened on Earth right back among the oldest traces of life. Exploiting other organisms is something we can expect on any planet where life exists.

Exactly what kind of organism it was that began eating these bacterial mats isn't clear. Certainly, very ancient fossils show winding tracks across the bacterial mats that presumably indicate someone was munching their way across the surface. For many years it was assumed that these strange tracks were made by ancient animals crawling over the living surface and grazing on the bacteria as they went. Perhaps they were, but no fossil of the animal itself was ever found, and so scientists assumed it was another one of the mysterious Ediacaran soft-bodied creatures without shells or bones that could not leave any kind of permanent record. Perhaps even an ancient ancestor of modern animals, like some sort of slug. But recently scientists came across remarkably similar tracks on the sandy floor of the Caribbean Sea around the Bahamas. Following these tracks led them to a remarkable discovery: a giant single-celled amoeba that looks rather like an oversized grape. Whether or not this is similar to the ancient mat-munching organism, it's a useful reminder that you don't have to be an animal to move, and that movement can be traced back to the most ancient and simplest of organisms.

Now simple geometry comes into play. Once something starts eating a stationary resource, it must learn to move. If the grazer eats faster than its food grows, it must move on in search of more food or die. Even if the food grows faster, the grazer will be so successful and have so many offspring that sooner or later – necessarily – competition with its children, and grand-children, and great-grandchildren will cause the food supply to become depleted. Someone has to leave home in search of their fortune. It's a cruelly simple law of evolution that energy is limited, and limited energy drives organisms to evolve ways to find new energy. Movement must arise. Aliens must move.

In case you are still not convinced that movement is inevitable, consider that another limited resource across the universe is space. As organisms reproduce, new individuals arise, and those individuals have physicality – they take up space. Even plants 'move', in the sense of dispersing their offspring to other locations. If no one moved, there would simply be no more space for new individuals, and evolution would cease. Life could continue for aeons with unchanging, immortal individuals, but they would never develop new traits, new abilities or new characteristics.

So organisms, wherever they live, must move to find space and to find energy. But it is the latter that drives the diversity of movement strategies that we see on Earth and, inevitably, must be mirrored on other planets. Space doesn't run away from you. Energy does. We've already seen how the animals of the Ediacaran period seemed to live in harmony with each other – more or less – and predation wasn't enough of a threat to cause the evolution of any kind of armoured protection such as shells or hard spines. There is little agreement among scientists as to whether such a situation could have continued indefinitely. Perhaps there was some environmental trigger, a change in ocean temperatures, or perhaps changing oxygen levels, that tempted one animal to take a bite out of another.

It is possible that a very specific set of circumstances is necessary for the evolution of complex animals that swim, bite, sting and hide.⋆ But an alternative interpretation is that such a development is inevitable, given enough time – possibly a very long time indeed, but eventually, hunters and the hunted must evolve.† The argument for this is that the idyllic Garden of Ediacara seems unstable, rather like a coin balanced on its edge. Yes, it could remain standing forever, but it really only takes the slightest disturbance, and then it falls – and there is no going back. In evolutionary terms, the choice is stark: if something is coming to get you, you will be eaten, and having a means by which to run away becomes very appealing.

The evolutionary drive to escape predators and the predator's drive to catch its prey are linked in a vicious circle, often called an arms race. An antelope that can outrun a cheetah will survive, but the cheetah will not. So predators must become faster, and then the antelope is under tremendous pressure to become faster itself, and so on and so on. Where does it all stop? Might there be an alien planet with supersonic predators and prey? Perhaps it never ends – will animals simply get faster and faster until they reach the speed of light? No, of course they will not.

⋆ See *Rare Earth: Why Complex Life is Uncommon in the Universe* by Peter D. Ward and Donald Brownlee.
† See *The Cosmic Zoo: Complex Life on Many Worlds* by Dirk Schulze-Makuch and William Bains.

One of the most fundamental rules of natural selection, on Earth and across the universe, is that there is always a cost–benefit trade-off. Improving your abilities in one field must reduce your capabilities in another. At the simplest level, energy is limited, and that energy can be used either to propel you more quickly or to build babies. One can imagine that in a world where cheetahs and antelopes use all their energy to move fast, an individual that moves a little slower but has more babies might be at an advantage. Eventually, other constraints always come into play and reduce the advantage given by excessive traits. If we do observe such excessive traits, it can only be because the balance of the trade-off is shifted very heavily to one side; for instance because the energy to propel the animal at such phenomenal speeds is incredibly cheap, or the threat of predators is unbelievably great.

The thought experiment of whether it would *ever* be possible for a supersonic antelope to evolve on another world illustrates another important feature of natural selection: that tangible benefits must accrue at every step of the evolutionary pathway. Reaching supersonic speeds is a particularly problematic challenge, because as you approach the speed of sound in whatever substance through which you're moving (typically air or water, on Earth), shockwaves are created, which dissipate a lot of the energy used by the animal. So before you can become supersonic, your movement becomes very inefficient – much of your effort goes into those shockwaves, rather than into making you move faster. Of course, human engineers figured this out, and realized that *if only* you could get past that sound barrier it would be worth the effort; and that's what Chuck Yeager did in the X-1 rocket plane in 1947. But in natural selection there's no 'if only', no foresight. If an animal doesn't gain a benefit from travelling at *nearly* the speed of sound, it will never evolve to travel faster.

On Earth, the speed of sound in air is about 340m/s – more than ten times the top speed of a cheetah, and still three times faster than a diving peregrine falcon (the fastest animal on the planet). But moving fast in air appears to be easy, relative to moving in water. The sailfish's 30m/s is comparable to the speed of a cheetah, but the speed of sound in water is a massive 1500m/s, so on this planet sea creatures still have a long way to go before they can become supersonic. The thicker the liquid or gas the more drag there is, and the less likely it is that an animal could become supersonic. Even on other planets, where life might exist in other liquids (methane, for instance), the chances of a supersonic creature seem slight. Jet propulsion through a gas – the only way that we know of to travel faster than sound – might also be the only possible evolutionary route for animals as well. But they would still be constrained to accrue a real benefit at every step of the way, every increase in speed. Otherwise, it just won't evolve.

Being a moving animal

To be confident that we have a comprehensive list of all the different ways that animals on other planets could move, we must be very systematic. We can use examples on Earth to see how adaptations evolved to overcome specific challenges. For those ways of moving that we do see on Earth – and that may very well cover the majority of possibilities – we can examine how these mechanisms evolved, what environmental factors caused them to confer an advantage on the animals that used them, and why they have evolved so far and no further. This gives us an excellent indication of what is possible on other planets and what conditions would make one type of motion advantageous over another. Fortunately, the constraints on how you can move are very simple and are tied to the laws of physics, which are of course the same on every planet.

Isaac Newton declared that there can be no acceleration

without force, and this simple fact lies at the heart of all animal motion. If you can't exert a force, you can't start moving. This is easy to believe if you watch a duck moving its legs frantically on an icy pond without making any progress, but is somewhat less clear when the duck spreads its wings and takes off into the air, seemingly without any concern for against what, exactly, it's exerting a force. In fact, that contrast between movement at a solid surface and movement through a sparse medium like air or water is the clue to how we can classify potential motion.

Let's deal first with some definitions and some basic distinctions. Space is either filled with something or it isn't. Let us put aside the question of motion through a vacuum and consider what might fill the space through which you want to move. You might want to move through something solid, as a mole or an earthworm does – we will address this later, as the question of whether moles and earthworms really move through a 'solid' is not quite as clear-cut as it sounds. But if the material filling space is not solid, it must be fluid. Most motion is through fluids, for obvious reasons (it's easier than moving through a solid). The term 'fluid' refers to anything that flows – it moves around you as you move through it, and this makes motion much more straightforward. Both liquids and gases are fluid, even if in common usage the word fluid is often reserved only for liquids, not gases. But gases certainly flow around you, and this is what is important for our purposes.

There are two main differences between a liquid and a gas. Firstly, liquids are usually (but not always) more viscous and denser than gases. They provide more resistance to movement, and this can be a disadvantage (slowing you down) or an advantage, because a liquid gives you something against which to push. Consider how easy it is to swim in water, but how much harder to move forwards simply by flapping your arms through the air! The other difference is that while gases tend to expand to fill whatever container they're in, liquids tend to accumulate at the bottom. There are some exceptions to this, but the principle is important, because it means that, most of the time, a liquid will have a top surface, and the interface between the surface of the ocean and the air above it has proven to be extremely important in the way that life has evolved on Earth.

If you are an animal living in a fluid like water, there are three possibilities: you might float, you might sink, or you might be exactly the right buoyancy to stay right where you are and neither float nor sink. If your density is higher than that of the fluid, you will generally sink, and unless you do something about it, you will eventually reach the bottom of the fluid. In that case, your challenge is movement at the interface between a solid and a fluid, and this is what happens with crabs, starfish and all the other creatures that live on the sea floor. The same goes for those animals that we see around us, dogs, cats, humans, moving at the interface between a solid (the ground) and a fluid (the air).

Of course, this assumes that the main force on the planet is that of gravity. There are good reasons to think that this is likely to be the case: for one thing, gravity is the only fundamental force that works well at normal distances. It is conceivable that on a small planet (therefore having a small gravitational pull) with a strong magnetic field, and lots of iron-based life forms, objects may be pulled in almost any direction. On such a planet, 'down' would not necessarily be 'towards the centre of the planet', as it is on Earth, keeping us tied to the surface. Rather, the magnetic field might pull creatures up at some places on the planet, and sideways at other places – there would be no absolute sense of up or down! However, the strength of magnetic field needed to tie our hypothetical life forms to the surface would be so strong that it might rip apart any complex molecules that formed. We can keep our minds open to bizarre possibilities, but nonetheless restrict our consideration to gravitational planets.

Some animals sink, but don't want to. The most obvious example is that of birds. When flying, they are moving through a fluid without sinking, but that is not their natural state – if they stop trying to keep themselves up, they will fall. There are many marine animals that are faced with the same problem – although because their density is much closer to that of water, and because water is more viscous than air, if they stop trying to stay afloat, the consequences are going to be much less severe than for birds. An octopus, for instance, generally walks across the seabed using its multiple legs, but can also, when needed, use jet propulsion to move quickly through the water, without touching the ground.

Notice that most species that are capable of fighting against gravity will do so only occasionally – it is likely to be more energetically demanding than letting physics take its course and moving along the lower interface where you belong. Anyone watching crows lazily feeding in a field knows how reluctant they appear to be to take to the air. However, that is not always the case. Some (mostly microscopic) animals spend their entire lives struggling to stay afloat, using tiny hairs that surround their bodies as multiple oars, constantly beating and rowing against the relentless force of gravity. Their small size means that the energy demands are less severe, and evidently the advantage of being in the water column, rather than on the sludge at the bottom, makes it all worthwhile. Smaller planets with lower gravity might find more and larger animals floating in a water column.

Some animals are less dense than the fluid in which they live, and so they float. On Earth, these are mostly air-breathing sea animals like seals, for whom it's an advantage to use less energy to come up to the surface than to dive down. For them, the world is upside down. Like birds in air, if they stop swimming they 'fall' (upwards) out of the water. Their motion is similar to the struggle of birds to stay aloft, although in this case the seals 'struggle' to force themselves downwards.

For animals that don't breathe air, floating would be just as inconvenient as falling out of the sky is for birds. But some of these animals float nonetheless. Bizarrely, the bluntnose sixgill shark seems to be positively buoyant, and the speculation is that they use their ability to float upwards silently so that they can sneak up on prey from below. There are, however, some non-air-breathing animals that have adapted to float because they gain an advantage from living at the interface between the water

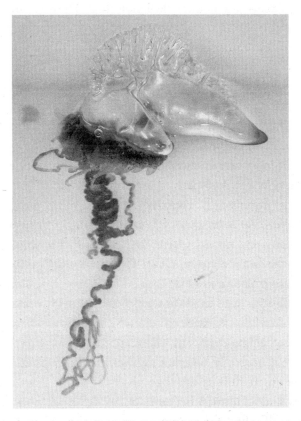

The Portuguese man o' war floats at the surface of the sea, using its gas-filled bladder. Lacking any means of independent motion, it drifts with currents, catching prey in its stinging tentacles.

and the air, and the Portuguese man o' war is the best-known example. This jellyfish-like creature uses a gas-filled bladder to remain at the surface, where it feeds on the small fish and plankton that congregate there. But the man o' war can only float. Like, say, an elephant, it is trapped at the interface between two substances, but unlike an elephant, it has no power of independent movement along that interface.

Many sea creatures, most familiarly fish, are neutrally buoyant – they make sure that they have the same density as the surrounding fluid, so that they neither float nor sink. They usually control their density using special tissues full of oil that are lighter than the surrounding fluid, or by having specialist gas-filled organs that compensate for their weighty body parts. These creatures are the most liberated in terms of motion. They can move forwards and backwards, limited only by the friction and resistance of the fluid through which they're moving. But they have an even more impressive ability; once they are moving, and the surrounding water is flowing over their bodies, they can redirect that flow, for example by using fins, to produce other motion: upwards, downwards, twisting around. Forward motion in a fluid opens the possibilities of moving in any direction, as well as twisting and turning. This fluid, dynamic motion is incredibly powerful, and also explains the acrobatic behaviour of bats and the murmurations of starlings, which use air flow rather than water flow, as well as all fish from anchovies to zebrafish, and dolphins, whose abilities to manoeuvre in water leave human observers feeling dizzy.

Once a neutrally buoyant animal moves through a fluid, it becomes incredibly sensitive to small variations in the direction of flow and its motion becomes unstable. Some means of stabilizing itself – fins, for example – must evolve. These in turn are easily co-opted to redirect that fluid flow and to provide manoeuvrability. It is almost inconceivable that only on Earth have animals discovered these tremendous advantages of fluid dynamic movement, and alien planets are probably swarming with, not necessarily fish, but animals that move like them.

Getting around in different fluid environments

Neutral buoyancy and motion through a single fluid

If you live surrounded by a single fluid – water, air, liquid methane – you need to exert a force against something that is not solid. Some kind of flapping or swimming motion will generate at least some force, and this is indeed what most animals do. By pushing fluid backwards, a forward force is generated; the principle is similar to that of a jet engine. But the precise mechanism by which animals swim through a fluid is unbelievably complicated and, surprisingly, often still not completely understood. Consider, for instance, a human in a swimming pool. If you push backwards you generate a forward force, but now what do you do? You must bring your hands forwards to make another stroke, but that would generate a backward force, and you'll end up going nowhere. Of course, you soon learn breaststroke, where by altering the shape of your hands you generate less backward force when bringing your hands forwards than forward force when making a stroke. Or you learn front crawl, where by bringing your hands forwards through the air rather than the water you generate less backward force.

Animal flight and swimming use similar techniques, altering the configuration of the force-generating machine (wing or fin, etc.) so that the forces don't cancel out. But the gains are often highly marginal. It is famously (but incorrectly) stated that according to the laws of physics bumblebees cannot fly, but as they don't know anything about the laws of physics, they fly anyway. In fact, most insects, birds and fish shouldn't be able to fly simply by flapping. But the fact is that they do, because they make use of a huge range of little fluid dynamic tricks to increase the amount of force they can exert. For example, most animals that use fluid dynamic motion also make use of vortices, the little whirlpools that you can see when you take a stroke in a swimming pool, and are a by-product of the force-generating movement. Vortices are made of fast-moving fluid, and they can be captured to give an extra little push forward. Many fish swim by moving their tail symmetrically from side to side with no change in configuration (unlike a human doing breaststroke). How can that produce a consistent forward force? Vortices hold the answer here too, but only with recent developments in imaging the motion of the water particles behind

Hummingbirds and fish are propelled forward by tiny vortices of spinning fluid (air or water) that accelerate a backward jet, which pushes the animal forward.

the fish have we understood just how important vortices are for much of animal motion.★ The swishing of the tail from side to side creates spinning rings of water, somewhat like smoke rings, that alternate in direction, and propel the fish forwards.

Moving through a fluid (either liquid or gas) seems like a likely scenario on alien worlds, but the challenges of generating a sufficiently strong net forward force are fundamental to the properties of fluids. Vortices will form in any fluid and, in the absence of any other way for bumblebees to fly, natural selection is just as likely to hit upon similar solutions on other planets. Alien bees will buzz like ours.

Apart from the obvious paddle-like motion of bird and insect wings and fish fins, there are other ways of moving through a fluid. As mentioned earlier, microscopic creatures can use a covering of tiny hairs called cilia that beat in coordination to propel them through the water. Mostly, this only works at the very smallest sizes, but the incredible comb jellies (among the most ancient of animal forms and not particularly closely related to the jellyfish) use an assemblage of rippling small hairs to propel them through the water at fairly modest speeds (1–2cm per second).

More dramatic still is the jet propulsion of squid, octopuses and the fossil-looking nautilus. By squirting out a jet of water backwards at high speed, they can generate a sudden forward force which is useful for escaping from predators. But while the nautilus uses jet propulsion regularly, this is only used as a last resort for squid and octopuses – jet propulsion seems to be inefficient compared to the rowing motions of fish and birds. Still, jet propulsion similar to that of the squid was used by their relatives, the ammonites, which were extremely common over a period of more than 300 million years. Under the right circumstances it

---

★ See a detailed description of this effect in Chapter Four of *Restless Creatures* by Matt Wilkinson.

A reconstruction of ancient ammonites.

would seem that squirting a jet of fluid is a perfectly reasonable way to get around, on any planet.

Fluids are rarely still. Most importantly, temperature differences, perhaps caused by the heating of the sun above, or warm rocks below, cause differences in density and pressure that make currents, which move the fluid from one place to another. Animals can abandon themselves and let the currents take them where they will, and many, many plankton and other sea creatures do exactly this. But inherent motion of the fluid medium has another effect: it can surprisingly be used to generate forces in other directions, and so power motion with very little effort on the part of the animal.

Birds are, as we know, denser than air, and so sink, potentially catastrophically. But by angling their wings just right, the flow of air can generate an upwards force, 'lift', that balances their own weight. They become magically neutrally buoyant, just like fish. This lift, generated by airflow, is the same force that keeps aeroplanes in the air – in case you're wondering (as most people do at some point) why aeroplane engines point backwards instead of downwards. Moving forwards generates airflow over the wings, which generates lift.

Maintaining buoyancy while gliding somewhat limits the flexibility that birds have to determine their own direction, but just as the flow of air can produce an upward force, birds can alter the shape of their wings to generate a force to the left or right as well, just as the pilot of a glider plane has a certain degree of control over where to go. Fish and insects go to great lengths and expend a lot of energy to generate fluid flow over their fins or wings, but an albatross takes advantage of the cease-less flow of the wind not just to float, but also to move.

My point here has been to emphasize that moving through a fluid is difficult – not because of the kinds of fluids here on Earth, but because such an insubstantial, flighty medium doesn't provide much on which to get a grip. On the other hand, the advantages of living in a fluid are tremendous: it provides much less of an obstacle to motion than a solid does. Animals, therefore, have found a diversity of ways of taking advantage of the fluid medium. Insects fly – barely, but spectacularly – dolphins twist and turn, jellyfish lazily paddle, and ammonites once squirted across the seas. Although we can't be sure that animals have exploited every pos-sible technique for moving through a fluid, there doesn't seem to be anything special about the fluids we have on Earth (water and air, mostly) that has led to strategies specific to those fluids. While we can't rule out the possibility that novel motion techniques exist on other worlds, we can be confident that at least some of the ones we see on Earth are there on other planets as well.

Achieving neutral buoyancy in water (or any other liquid) is considerably easier than in air, or any other gas. Air is about 1,000 times less dense than water, and so there are some important differences between liquids and gases. Very few solid materials will float in air, and so exploiting the atmosphere has proven much more challenging for animal life than exploiting the sea. In theory, it is possible to imagine an animal floating through the air under a sack of gas, most likely hydrogen, which is generated anyway as part of the metabolism of many different kinds of bacteria and other microbes. This could be an incredibly efficient way to travel around the world, feeding on any 'atmospheric plankton', in a similar way to how blue whales feed on massive quantities of krill in the ocean. Rather than expending vast amounts of energy like swallows and bats in chasing prey that use powered flight to escape from you, a sky-whale could float without effort, harvesting microscopic creatures that float passively as it does.

Such an imaginary creature has been dubbed a Fortean bladder,* or just a 'floater',† but on Earth, no such creature exists. Why not, and might Fortean bladders be common on other worlds? Apart from the danger of catastrophic combustion, which spelled the end of the Hindenburg Zeppelin, and with it the era of hydrogen-based air travel for humans, there must be some reason why evolution has not led animals down this particular path. Viscosity holds the answer. In water, small creatures barely sink, even if they are denser than water. Currents and eddies tend to keep them mixed and in suspension, and even the feeble beating of their cilia is enough to suspend their tiny bodies in the water column. Not so in the air. Even a microscopic organism would be hard-pressed to stay afloat in a medium as sparse as the atmosphere. Cilia would be wholly inadequate, and only the flow and movement of air currents – admittedly often very strong – would keep the creatures airborne. The fact is that, on Earth, there are no microscopic aerial organisms to act as a planktonic food supply once you go up much above ground level, and so no floating whales to feed on them.

---

* See *Life's Solution* by Simon Conway Morris.
† See *Cosmos: The Story of Cosmic Evolution, Science and Civilisation* by Carl Sagan.

But other planetary environments could be more favourable to Fortean bladders or sky-whales. In a denser atmosphere, say on a gas giant like Jupiter, or on a smaller planet where gravity is weaker than on Earth, microscopic organisms could remain afloat for long enough for a whole food chain and ecosystem to evolve around them. Carrying this thought experiment further leads to other problems, however. A small planet with weaker gravity would have difficulty holding on to its atmosphere, which would likely escape into space. Mars, which has one third of the gravitational attraction of Earth, has an atmosphere 200 times thinner. We currently have a fairly limited understanding of the behaviour of the atmospheres on gas planets like Jupiter, but from what we can see they are violent and turbulent places, unfavourable for the evolution of life.

Some people have speculated that the clouds of Venus might host microscopic life, as I mentioned in Chapter Two. For a complete ecosystem to develop around such aerial plankton, including large planktonic-feeding sky-whales, organisms must evolve to be larger and larger, without simultaneously falling out of the sky. Neutral buoyancy is easy to maintain in liquid as animals evolve to be larger, but in a gas a tendency towards larger organisms would have to be matched, step by step, with buoyancy organs (hydrogen-filled, for instance). Unlikely, but by no means impossible. If, however, life does exist in such places, it may be best to look for analogies in the planktonic feeders of Earth's oceans, rather than in our creatures of the sky.

Movement at the interface between solid and fluid

We humans are stuck on the ground, along with caterpillars and elephants, but we manage just fine, and some of our relatives, like cheetahs and ostriches, seem to be positively expert at spectacular motion at an interface. The way that animals move on a solid surface is arguably the most important mode of motion on this planet, because it is likely that the first cells, and possibly life itself, began at the interface between solid and liquid.★ The earliest life for which we have hard evidence, the stromatolites mentioned above, formed on a solid surface, and the possibly single-celled grazers that munched on them moved across that surface, covered by fluid. The oldest forms of life are unquestionably linked very closely to this mode of movement. Gravity brings you down, and down eventually means the floor. Alien planets almost certainly have floors. So how to get around on a solid surface?

Generating a force to push you forward is much easier when you are up against a solid surface than when you are suspended in a fluid, but the laws of physics are going to throw you a vicious catch-22. You can only push yourself along a surface if there is friction, but friction slows you down. Anyone who has tried ice skating is familiar with this: skaters can move incredibly fast — but for the most part, we beginners flail around helplessly, unable to move at all. The first moving creatures may have been similar to single-celled amoeba-like organisms, slithering along the surface by extruding a part of their body forward, and then dragging the rest of the cell up from behind. The key thing about this kind of motion is that the whole cell is in contact with the surface all the time. There's little chance of slipping, but conversely you waste a lot of energy pulling against the friction. This is easily tested at home: lie flat on the carpet and try reaching forwards and pulling your body along — without lifting any part of your body off the floor. It's hard, and slow. Freeing yourself from that

---

★ One theory of the origin of life has a vital role for chemical reactions taking place on the mineral surface of exposed rocks, bathed in a salty liquid. The image of fatty deposits blistering off mineral surfaces has given rise to the picturesque term 'prebiotic pizza'.

friction while continuing to use it to get traction means being in contact with the floor only very slightly, lifting yourself onto your legs, for example.

Legs – the ability to minimize contact with the surface, while still being able to push against the surface – are such a phenomenally useful adaptation, it is hard to imagine a world in which they did not evolve at all. Of all the animals on Earth that live on a surface (on land or on the seabed), a substantial minority (notably the molluscs like snails and slugs) still manage without any legs at all, but the overwhelming majority lift themselves off the frictional surface. Molluscs may well exploit a niche where legs are not necessary – by excreting a slippery slime, they solve the friction problem in a unique way – but snails are notoriously slow, and while alien planets may have slug-equivalents, it seems unlikely they could be the dominant life form. Legged creatures would have a speed advantage over slime-creatures in almost any environment. Snakes, too, are legless, although they evolved from legged lizard ancestors, adapting to a life of digging underground. Legs are definitely an adaptation for life at the interface between solid and fluid. They just get in the way if you live underground or floating in the water.

Legs became most prominent in the arthropods, the best-known examples of which are insects, spiders and crabs. Quite independently, a very different kind of leg evolved in vertebrate fishes, which adapted their fins (used for propelling themselves through the water) to lift themselves off the ground. Many fish, such as the epaulette shark, still do this on the sea floor today.* But crucially, human legs are built completely differently from spider legs. Arthropods have a rigid exoskeleton, with their soft bits inside; vertebrates have rigid bones, with soft tissues outside. The two evolutionary pathways came up with functionally

---

* See *Restless Creatures* by Matt Wilkinson.

The epaulette shark walks along the sea floor on long leg-like fins. Fish fins like these were the ancestors of the legs we use – which are very different from those of spiders or beetles.

similar solutions to the problems of moving at a surface, using completely different mechanisms. Which is better? Neither, nor does it matter. Each innovation was based on, *and constrained by*, the details of the body plan of their ancestors. Vertebrate legs are the way they are because fish have bones, not because they are the right way to get a cheetah up to 100km/h. Arthropod legs are the way they are because an exoskeleton is a good way to avoid drying out on land, and their ability not to dry out was the key to the phenomenal success of the arthropods.

The important insight this brings is that life on alien planets is very likely to have legs – but not necessarily legs built as we know them. *Their* leg structure will be constrained by *their* evolutionary history too. If life exists on a world with a solid–fluid interface (and not just in a bottomless ocean), it seems inevitable legs will exist. But they will probably have evolved from a range of different starting points, and ended up with a range of different solutions.

Apart from arthropod and vertebrate legs, we find two other solutions on Earth (there are some more, like octopus legs, but these function primarily to manipulate food and other objects, rather than to lift the body off the ground). Velvet worms (so-called despite the fact that they are not particularly closely related to worms) have a row of peculiar stub-shaped legs on each side of the body. Unlike arthropods and vertebrates, the legs have no rigid components, and are basically fluid-filled bulges that can be moved by alternately stretching and lengthening different segments of the body. The motion is remarkably worm-like, but instead of being in contact with the floor along its whole body, only the pads of the feet touch the ground, and this makes them silent and effective predators. This mode of walking has been around for a long time, and the ancient ancestors of the velvet worm were possibly some of the most bizarre and surprising fossils that have ever been discovered. *Hallucigenia* lived more than 500 million years ago, and the delicately preserved fossils were so difficult to interpret that the animal, with

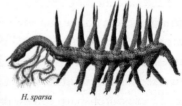

H. sparsa

A modern velvet worm (left) with its stubby, fluid-filled legs, and an artist's impression of the extinct *Hallucigenia* (right), which is thought to have moved around the sea floor in a similar way.

what certainly appears to be both a set of legs and a set of spines, was originally reconstructed upside down.★ But it certainly seems clear that this alien-looking creature walked over the sea-bed on some very unusual legs.

The fourth type of Earthling leg is that of the echinoderms – starfish and sea urchins – and this innovation is more bizarre still than even *Hallucigenia*. Starfish and their relatives have a rigid shell covered with tiny pores. By contracting muscles and by opening and closing valves, they can cause their internal water pressure to push parts of their bodies out through the pores, forming short stubby 'legs', called tube feet. This multitude of legs lifts the animal off the floor, and by coordinated extension and contraction of the different extrusions, it can move along the seabed. Interestingly, the legs aren't synchronized in their motion, as we see in just about every other form of leg-based locomotion (horses, beetles, even velvet worms). Instead the chaotic-looking multitude of moving pieces still manages to get the animal to where it's going. Echinoderm legs are an import-ant reminder that the conventional evolutionary solutions with which we are most familiar (often those that we ourselves use) represent only a small part of all the possibilities. Far from need-ing to speculate wildly, it only requires a closer look at the diversity of solutions we already have on this planet to get ideas for possible alien ecosystems.

So what kind of planet might support life forms that predom-inantly use legs unlike our own? Echinoderm motion is impressive, but slow – sea urchins typically move only a few centimetres a day. Clearly, a world of fast predators is not likely to engender a diverse range of animals with tube feet. Indeed, most sea urchins have a crown of sharp spines, so that they aren't easy targets for fast-moving fish (restricted to the fluid medium), or octopuses and

★ See *Wonderful Life* by Stephen J. Gould.

The tube feet of a sunflower sea star, *Pycnopodia*.

other predators (on the sea floor). One possible advantage of a multitude of tube feet is if negotiating over a rugged and difficult surface. Over particularly creviced and rocky ground, being able to 'step' on whatever surface presents itself could be a distinct advantage. Sharp spikes might puncture a large foot, but multiple tiny feet could safely navigate even broken glass.

More importantly, a surface with very low friction is also best served by a large number of legs and tube feet. On a slippery surface, a leg can only get enough traction to provide the smallest of forces. Many small legs will add up to enough force to get the animal moving. Add to this a particularly viscous fluid above the surface, and traditional leg-based animals would seem at a substantial disadvantage compared to our hypothetical sea urchin-like alien. It would be like trying to walk in a non-stick frying pan filled with oil, as the thicker the liquid, the more force it exerts back when an animal attempts to accelerate.

To sum all this up, legs are almost essential in any surface eco-system. They reduce friction and so increase the speed of an animal, and speed is of the essence when trying to catch prey or evade predators. That limited resource, time (manifested here as speed), is as potent a force as space or energy. However, exactly what form legs will take depends both on the properties of the solid surface (smooth or jagged, low or high friction) and of the fluid above (runny or viscous). Fortunately, we have enough of a diversity of adaptations here on Earth to give us at least potential mechanisms that seem appropriate solutions even on worlds almost unimaginably different from ours.

Life in the underground

Finally, we must consider those creatures that move through solid ground. Moving through a solid sounds impossible, and in a sense it is, as a solid cannot flow around you like a fluid. But moles and earthworms, and a few other animals like those that burrow into the seabed, still manage to live and move through what appears to be a solid medium. In reality, though, at the scale of these tiny animals, the earth is not really a solid, but a collection of solid particles with quite a lot of fluid around them. Subterranean animals mostly move simply by pushing solid par-ticles out of the way, or by taking bits of soil from in front of them and placing them behind – sometimes by ingesting them at the front and excreting them at the back. Darwin was fascin-ated by everything to do with earthworms, and carefully observed their mechanisms of movement:

Means by which worms excavate their burrows.—This is effected in two ways; by pushing away the earth on all sides, and by swal-lowing it. In the former case, the worm inserts the stretched out and attenuated anterior extremity of its body into any little crevice, or hole; and then, as Perrier remarks, the pharynx is pushed forwards into this part, which consequently swells and pushes away the earth on all sides.★

As well as rhythmical undulations in earthworms – gripping the surrounding soil in the front part of the body and pulling the back part forward – other animals such as moles simply press the soil into the sides of a tunnel with large paws. This is only possible because the earth is not really solid at all but filled with air. Occasionally there are animals that genuinely bore into solid rock, either by rasping away at it – such as the unfortunately named 'boring clams' or 'piddocks' – or even dissolving the rock with acid, as another marine bivalve does, the 'date mussel' *Lithophaga*.

However, it is difficult to imagine that an entire ecosystem could comprise underground creatures. We believe – with tentative confidence – that liquid is essential for life. Chemical reactions barely take place in solids or gases, and so life must require liquid of some sort. Even if a planet has much of its life underground, it probably first evolved, and diversified, in fluids. Certainly on Earth underground animal life is quite limited, and most animals that do live underground retain an essential connection with the fluid world. Meerkats come above ground to forage, and clams extend a 'siphon' through the sand to access the seawater. An alien world with an extensive underground ecosystem would be a huge surprise, but a rare opportunity to study the constraints of life under conditions that, perhaps, go beyond what we can predict from the diversity of life on our own planet.

★ *The Formation of Vegetable Mould, Through the Action of Worms, with Observations on Their Habits* by Charles Darwin.

Fearful symmetry: shape and movement

Up until now we have skirted around one of the most dominant and important movement strategies on Earth. Something that is so ubiquitous that we take it for granted without thinking twice. It's a feature of almost all life on Earth, so familiar that we don't even stop to think whether or not we would expect this on alien planets. But all this talk of multitudinous independent tube feet leads us to one of the most important questions of all: will aliens be omnidirectional like a sea urchin, i.e. able to move in any direction without preference? Or will they be symmetrical like us, with left and right sides, and consequently a front and back and preferred direction of travel?

Almost all modern animals, with the notable exception of sponges, jellyfish, comb jellies and corals, have a left side and a right side – more than 99 per cent of the species alive today. We say that they are *bilaterally symmetric* around their middle. It's hard to overemphasize how important this body plan has been on Earth for the evolution of animal movement. Bilateral symmetry can be an advantage if you need to crawl like an inchworm along a solid surface, lifting up part of your body, and placing it down further along. Being able to place your body 'further along' implies that your body is elongated along some axis, and so has a left and a right on either side of that long axis. The first wriggling creature was likely similar to modern flatworms.

The advantages of bilateral symmetry to animal movement are tremendous; so much so that hardly any animals remain that do not use this approach. Having a front and a back means you know which way you are going, and your movement organs (such as legs) can specialize in getting you in that direction. A left–right symmetry also opens the possibility of particularly effective locomotion using a mesmerizing wave-like motion of the appendages.

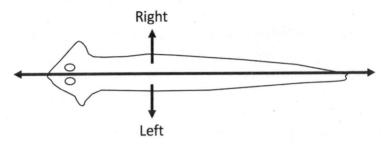

A modern flatworm (planarian), possibly similar to the first creatures that began to move in a directed way. Having an elongated body that allows the animal to inch forwards also automatically gives it a left side and a right side.

Manta rays and squid undulate their fins, and millipedes glide along on waves of motion of their tiny feet. No animal that lacks this symmetry can compete with the speed and energy efficiency of the left–right appendages, whether legs or just flaps of skin.

But understanding the exceptions to this rule is absolutely essential in deciding what possibilities may exist on other planets, and whether other planets may host life that is completely different in its physical organization from what we see around us. The remaining few species that don't belong to the bilateral group are mostly corals and jellyfish. Corals don't move at all, whereas jellyfish swim lazily through the open ocean feeding on plankton and other small creatures. It's not immediately clear why jellyfish would not benefit from a bit of symmetry. Certainly, they are faced with dangerous and (relatively) swift predators like sea turtles, and you might have thought that a jellyfish that could use a left-to-right undulation to swim faster would stand a better chance of escaping. But clearly in the niche that jellyfish occupy, the advantages of evading predators with a more sophisticated mode of movement simply don't outweigh the advantages of retaining such a simple body plan and lifestyle. One of the main drivers of the diversity of life is that, if most are the same, being different often presents its own advantages.

Interestingly, a few bilateral species have lost their body symmetry and gone back to a simpler body plan, presumably because there's something about the environment in which they live that makes this geometric organization more advantageous. In such cases we can learn a lot about the conditions that may or may not favour a bilateral body plan on other planets. Those renegades are our old friends, the tube-footed echinoderms, who have abandoned their bilateral habit in favour of a radial star-shaped symmetry. We know that starfish and sea urchins are in fact bilaterally symmetrical in their origin because their tiny free-swimming larvae look very much like they have left and right sides. But when they find a suitable place to settle and they transform into the adult animal, one half of their body wastes away and they become quite prominently circular or star shaped. Why would they do this?

The answer is probably tied to their tube feet. In an environment where tube feet are an advantage – perhaps an irregular, sharp terrain, perhaps a low-friction surface in a high-viscosity fluid – bilateral symmetry becomes less important. When supported by a mass of independent feet, directionality seems meaningless. Echinoderms and, I suspect, their alien parallels move cautiously on thousands of tiny feet, seeking out sufficient traction, where a fast-moving bilateral would slip and fall constantly.

And aliens? Movement on Titan and Enceladus

Now that we have surveyed the evolutionary demands on movement, how can we make predictions about aliens, and test them? Two of the moons in our own solar system that are the strongest candidates for supporting alien life tantalizingly have large bodies of fluid in which that life might exist. Saturn's moon Enceladus has a saltwater ocean totally enclosed under more than 30km of ice. Recent analysis of measurements from the Cassini space probe shows that this ocean extends around the entire moon, and could be as deep as 30km itself (for comparison, the deepest part of the Earth's ocean, the Mariana Trench, is about 11km deep, but most of the Earth's seas are no more than 3 or 4km deep). The pressure and temperature differences between the Enceladus ocean floor and the ice cap above must be tremendous; it seems unlikely that organisms traverse that whole range.

Any life on Enceladus either occupies the ocean's boundary with the upper ice cap, the ocean floor, or swims through the ocean itself. If the latter, there seems to be no reason why the strategies evolved on Earth – paddling, squirting and rippling cilia – should not work perfectly well on Enceladus too. But at the boundary between the water and the ice another ecosystem may be possible. Unlike the sea floor on which life evolved on Earth, on Enceladus the solid floor is *above* the liquid sea. Creatures exploiting that niche would need to be positively buoyant, floating up in the water, rather than sinking down. Their buoyancy would overwhelm and effectively take the place of gravity, so that their 'weight' would pull them upwards.

In many ways, the entire ecosystem would be similar to a sea floor, but upside down. There seems no reason why all the complexities of an Earth ocean floor community could not exist the other way up, between the water and ice on Enceladus. The alien equivalents of crabs scuttling across the underside of the ice, worms and other soft-bodied creatures burrowing up into the ice for protection, predators with eyes on their upper surface, swimming through the water, looking for creatures clinging to the ice upon which they could swoop *up* and catch. Even on Earth, the undersides of icebergs provide a simple ecosystem of algae and algal grazers. The main problem with such an ecosystem though is that dead animals will most likely sink (solids being denser than water), depriving this ecosystem of one of the main sources of food that fuels life on the sea floor. Rather than a rain of organic matter from dead plankton, the dead might 'rise up' from the floor and disappear out of reach into the depths!

Another moon of Saturn, Titan, has surface lakes and rivers, with rain eroding mountains as it flows down to the sea. However, it is much too cold for liquid water to exist – the surface temperature is about -180°C. The liquids on Titan are hydrocarbons that would be gases on Earth, like methane and ethane. Radar measurements from the Cassini probe showed that, while not as unfathomably deep as the subsurface ocean of Enceladus, at least some lakes on Titan are as deep as 160m, comparable with Loch Ness in Scotland. If life exists in these lakes, it will be based on a biochemistry so different from life on Earth that it is almost impossible to describe accurately.

However, the mechanisms that such life would use for manoeuvring around could be little different from those used by the animals of Loch Ness. Much is still unknown about the fluid properties of Titan's lakes, but they are thought to consist mostly of liquid methane, which is about six times runnier than water. So in a very methane-rich lake, movement strategies may be more like those we see in air on Earth, rather than in water. The effectiveness of paddle-based limb movements would be much less, and moving small cilia through a runny liquid may not even be enough to keep small organisms suspended in the (methane) column.

On the other hand, careful observations of reflections from the surface show that there do not seem to be any waves larger than a few millimetres. Perhaps there is very little wind on Titan, or perhaps the lakes are thicker and more viscous than we expect. Experiments at the Jet Propulsion Laboratory at Caltech have simulated the conditions on Titan and proposed that benzene could fall as snow and dissolve in the lakes to produce a thick saturated solution, rather like the salt in the Dead Sea in Israel. In such an environment, perhaps cilia or jet-based propulsion would be more advantageous, and the soupy, sticky nature of the fluid would strongly favour plodding, almost burrowing movement strategies; perhaps even the evolution of tube feet. But even if the relative advantages of the different mechanisms are different, the repertoire of movement techniques remains the same.

We can't be sure that our survey of movement strategies on Earth has been totally comprehensive of all strategies possible throughout the universe. But by beginning with the physical properties of motion, we have seen that pretty much every mechanism we can think of has been exploited here on Earth. If the physical environment of an alien planet in any way resembles that of Earth, we can be confident that at least some of the movements of alien animals will seem very familiar to us indeed. There must exist very odd planets with odd creatures moving in odd ways, but the majority of alien worlds will feature similar constraints to Earth and so similar solutions. The mechanics of motion is too straightforward, too consistent, to allow utterly bizarre movement strategies to evolve. Life in the universe will basically move the way that life on Earth does.

Of course, it is possible that alien creatures will be physically utterly different from what we see around us. Perhaps they can move effortlessly through solid rock because they, themselves, are not solid. Gaseous aliens, perhaps. We cannot discount such suggestions completely, but we think it is unlikely, because the nature of life appears to involve concentrating energy in a single place, rather than letting it spread out and become diluted. But even if gaseous aliens are a possibility, we can be confident that they are rare, compared to those membrane-bound creatures like us that have to choose from a simple set of movement strategies, just like life on Earth does.

As a result, we can be confident that *most* alien animals will be bilaterally symmetrical. Many of those that live at the interface between solid and fluid will have legs, and those legs will probably look familiar to us. Aliens living in a thin fluid like air will have to float like balloons or use air flow to generate lift and stop them from sinking. Those in a denser fluid like water may be neutrally buoyant, but will generate forward movement by paddling, undulating or using jets of fluid like a nautilus. How extraordinary that an alien landscape will likely be instantaneously recognizable to us, just by the way that its inhabitants move around their world.

5.   Communication Channels

Take a walk in the woods and you will be surrounded by animals: animals chirping, rustling and screeching. You can't help but be aware of the sounds of myriad creatures all around you. But you will also sense these animals in many different ways: you see the blackbird hopping along the ground with his tail in the air; become irritated at the gnat that brushes your skin; and perhaps you may smell a family of foxes or, more likely, a pile of horse droppings from a nearby bridleway. It is obvious that we have more than one sense – sight, hearing, smell, touch and taste – but we rarely think of how those different senses give us complementary information about the world around us, and in particular about the other animals busily getting on with their lives. Some of these signs and signals in each of the different senses are directed at us or at other animals, and some are merely the by-product of normal animal activity: the buzzing of a bee's wings as it flies, or the shower of beech nuts as a squirrel races through the trees.

But many clues to the presence and activity of different animals are, in fact, communication; they have evolved to suit the purposes of the animal sending the signal. This communication is carried over different sensory channels, like sound and light, that we will call *modalities*. Oftentimes signals★ are created in different modalities by the same activity (perhaps, a rabbit digging

---

★ Strictly speaking, scientists use the term 'signals' only for something evolved for the purpose of communication. The sound of a rabbit digging would be called a 'cue'. However, this is so contrary to popular use, that I choose the simpler 'signal'.

a scratching in a bank), radiating away from the animal like ripples on a pond, simultaneously as sight, sound and (although perhaps not detectable by our own human noses) smell.

The communicative modalities

Why do we have so many senses? Is it really necessary that we (and other animals) can simultaneously see and hear what is going on around us? There are some good reasons to think that using multiple modalities to sense the world and also to communicate gives animals a robustness in the face of changes in the conditions around them. On a stormy day, heavy rain may drown out sound, but at night when all is dark, sound may be the only way to sense the environment. This is the property of *multimodal robustness*, where sending the same message through different channels helps to ensure that the message is less likely to get lost.

On top of that, using multiple modalities can sometimes help animals to convey a richer set of information than using just a single modality: *multimodal enhancement*. Your dog can convey a great deal of information in his bark; in fact, studies have shown that humans are acutely aware of subtle differences in dog barking, and can distinguish between friendly, angry and lonely barks, even in the absence of any other cues. However, combine the sounds that a dog makes together with his body language – posture, tail wagging or held stiff, the direction of his gaze and the alertness of his ears – and these add tremendous additional detail to our perception of the dog's internal mental state, allowing us to understand a great deal about how he's feeling, even in the absence of a clear shared language. Dogs, as with their wolf relatives, have evolved a complex multimodal network of signals that provide their pack mates (whether canine or human) with detailed information about the social environment and how behavioural interactions between individuals are likely to proceed: will she attack? Will she run away? Will she mate with me?

So the world is full of signals being transmitted through different physical modalities, of which sight and sound are the most obvious to us. Is it a coincidence that our own language evolved as an acoustic signal? Are we unusual in using sound to communicate, or is there something special about sound itself that makes it almost inevitable as a signalling modality? Is the fact that we seem to be surrounded by acoustic signals just a peculiarity of the conditions on our planet? Or might there be something special about sound waves that makes them particularly suitable for being crafted as communication – and even into a language? Can we expect aliens to have a *spoken* language? If so, what would that tell us about the conditions on their planet, and if not, what kind of modality would evolve to be suitable for those conditions?

We can, of course, speculate with amusing thought experiments, based perhaps on some of the creative alien species we find in science fiction, such as the telepathy of Vulcans in *Star Trek*, or the circular symbol language of the heptapods in the movie *Arrival*. However, we can do better than that. We can rely on the same laws of physics that apply equally on Earth and on the fictional planet Vulcan, to consider which signal transmission modalities would work well for communication and which wouldn't, given the particular physical properties of the environment in which an alien lives. By going back to first principles, we can ask, what is absolutely essential for communication?

Perhaps the simplest definition of communication is that useful information is produced by one individual and transmitted to another individual, who then decodes that information. These are basic requirements, and surely universal ones, required even of the most bizarre and unexpected forms of communication on the most alien and unfamiliar worlds. It is these fundamental principles that cause scientists to dismiss the idea of telepathy – not because it isn't useful, but because we cannot see how a telepathic signal could be produced, transmitted and decoded. Should we find a way to explain telepathic communication that does satisfy the underlying physical laws, it would surely be worthy of investigation. I will return to this tantalizing possibility later in the chapter.

The idea that a signal contains useful information is particularly interesting, because some modalities are going to be better suited to conveying large amounts of useful information than others. Imagine sitting in silence in a classroom where your only way to communicate with the teacher is to raise your hand – just raising your hand, not talking or writing on the board (and, of course, not using sign language). While the teacher may be able to get a limited amount of information from the pupils (asking yes/no questions, for instance), you can see that this kind of binary on/off signal is a bit of a dead end as communication goes. If we want to know what kind of communication modality might support the eventual evolution of language, we can ask as a quick and rough rule of thumb, 'Could you write poetry like that?' It doesn't seem that our hand-waving classroom is a convincing candidate for poetic composition. The modalities that can support complex communication, and eventually language, need to be rich, nuanced, with the possibility of including large amounts of information in the signal.

Some signals can be transmitted by mechanisms that operate only at very short range. Touch is a hugely important sense to many animals; grooming and preening is a way of forming social bonds in primates and birds, and even humans receive detailed information about others from the way they touch you – firmly, casually, affectionately, etc. The whiskers of mice and moles allow them to find their way in the dark exclusively by touch. But these are very short-range modalities – you really need to be right next to someone to understand their communication, and that puts a big constraint on how useful they can be for complex signalling, and in this chapter I'm mostly going to concentrate on signals that can reach far and wide. Interestingly, though, there is at least one kind of touch signal that *is* long range. The whiskers of seals respond to tiny currents of moving water, and so can detect motion at a distance. However, this detection of vibration in a fluid environment really shares a lot of similarity with sound transmission and reception. It's just worth remembering that what we consider 'hearing' on this planet may be implemented quite differently on an alien planet – say by the sensitive whiskers of alien seals.

Consider a pack of wolves that need to cooperate to survive. The environment is harsh, food is hard to come by, and the only way to provide enough energy is to hunt infrequently but to hunt large prey that will sustain the animals with more than just a mouthful. To hunt animals larger than themselves the wolves must cooperate, and cooperation requires some form of communication. Wolves are a particularly interesting species for our investigation because they share many properties with our own ancestors: they must cooperate to find resources and defend themselves from other animals; they are highly intelligent, with the social skills to live in a large group of individuals; and, of course, they are very vocal. The nature of their communication – and that of our ancestors – must suit the requirements of that way of life.

The communication must be fast – there's no point trying to coordinate a hunt if the opportunities have passed before the signals have reached their intended recipient. It would also be useful if the signals can be localized – if you can tell who it is that is talking. It also seems important that a signal can be perceived without too much dependence on where the receiver is actually located – for example, we can hear sounds even if we are hiding behind a bush, but visual signals are restricted to being in the signaller's line of sight. Without wanting to generalize too much, and possibly assume more than we should about communication on an alien planet, some physical properties will be important if a signal has the potential to evolve and gain more complexity, and some modalities will possess those properties, whereas others will not.

Bearing all this in mind, how many different modalities are there, and how many of them can we directly observe being used on our own planet? It turns out that animals on our planet have evolved the ability to communicate using just about every modality we can think of. Some of them are very familiar – sound, vision, smell – others are quite surprising, such as the electric-field communication of certain fish. Even a magnetic perception is widespread among animals on this planet, although as far as we know, none of them use this sense to communicate directly with each other. Radio waves are not used either, but both this, and the possibilities of a magnetic modality, cannot be ruled out for alien communication. So let us take a tour of the dizzying range of animal communication modalities, and consider what those tell us both about the nature of communication and about the possibilities that may be available to animals on other planets.

Sound: our talking modality

Sound is how we communicate with each other. Yes, I am writing these words via a visual modality, but writing did not arise until hundreds of thousands of years after language itself came into existence. We also experience animal communication overwhelmingly through their sounds, and that can tend to colour our expectations of what animal (and alien) communication must be like. Of course, part of the reason why we think of animals making sounds is that, much of the time, we are unable to see them. Even when you hear a pigeon cooing or a cricket chirping, unless you take the time to look closely, you probably will not see the animal itself. This is no coincidence. Sound has a very important property that (on our planet) undoubtedly propelled it to be the dominant modality for communication. Sound travels around things. No matter that the pigeon is hidden in the leaves or the cricket in the grass, the sound reaches us anyway. Light is blocked by most solid objects; sound goes around them. The physics of why this is the case is not trivial, but in general it has to do with the wavelength of the signal: sound typically has wavelengths of around one metre; light a wavelength of about one ten-millionth of a metre. That means that sound waves barely notice smaller objects in their path, but slip around leaves and grass and trees, in much the same way as we ourselves might navigate obstacles in that same woodland. Light, however, on its microscopic scale, finds itself up against gargantuan obstacles, where every little molecule appears as a mountain to be overcome.

Obviously this picturesque description becomes completely different on a planet where scales are radically unlike those on Earth. In a hypothetical world smoother than a ball bearing, where complex life is microscopic, one could consider that the relative advantage to acoustic communication might be less. But for most ecosystems that we can imagine, light definitely scores fewer points than sound. Sound, however, has one very important drawback, which is that it can only be transmitted through a physical medium like air, or water, or soil. Light, on the other hand, can travel through the vacuum of space; it is present on the surface of the Moon, which is utterly silent. Similarly, on a planet with a very thin atmosphere – Mars, for instance – sound barely propagates at all. Although the atmosphere on Mars was once much thicker, today no creatures on that planet would be able to communicate effectively using sound: no one could hear you scream.

The second advantage of sound is that it is fast. Not as fast as light (not nearly), but so fast as to make little difference at the scale of Earthly organisms. When a signal travels at 340 metres per second, a signal sent acoustically from one animal to another arrives almost instantaneously. Few animals communicate with each other over distances greater than a couple of kilometres (leading to a delay of a few seconds), and those that do cannot expect an immediate response – their partner in conversation is too far away to do anything immediately anyway, as even the very fastest animals are still much slower than their sounds. Sometimes the delay in the transit time is more substantial, as with the songs of whales, which can travel hundreds of kilometres underwater, but even then, sound travels much faster in water than in air and the delay is still only measured in minutes.

Whatever the medium, speed is absolutely essential in the evolution of complex communication in complex social animals. Conditions change rapidly: the caribou swerves left and all the members of the pack must know to follow; the leopard is about to pounce on your mate and you must warn him. Complex communication provides help in solving difficult problems, and these problems are almost always time critical. Of course, we can speculate about worlds where the speed of sound is particularly slow, rendering it useless, or fast, possibly making it redundant if the pace of life is different. Consider an ecosystem inside a thick, soupy tar, where predators creep towards you more slowly than snails, and sound travels so fast that it's a waste of energy to use that modality when a slower signal will suffice equally well. We have to be aware of our assumptions and consider the extreme alternatives, but also consider that on many exoplanet environments sound will still be a tremendously useful way to communicate.

The other big advantage of sound is that it can convey a large amount of information in a very efficient and compact way – we say that it has a large *bandwidth* compared to, for example, our hypothetical classroom of hand-raising students who can only say yes or no. Sound is particularly good at this because, on the scale of our planet, and at the sizes of animals that live here, it is relatively easy to distinguish many different frequencies even when they are mixed together in an acoustic signal. Consider, for instance, how it is possible to understand one person talking even in a room crowded with people each engaged in their own separate conversation.

Here's a very simple yet technical experiment you can do yourself. Go outside one morning when birds are singing their dawn chorus and record a minute of sound on your phone. Then upload the sound file to one of the many websites that generate a visual representation of the sound called a *spectrogram*.★ You'll get something like the image opposite, which shows time from left to right, and frequency (pitch) from bottom (low pitch) to top (high pitch). The areas that are darker indicate that there is more of that frequency present – you can think of it as being similar to sheet musical notation. In this clip, recorded during the dawn chorus in England, there are at least four different species of birds singing. You can see each of them quite distinctly as separate shapes on the spectrogram, showing the subtle differences in the way that the birds modify acoustic frequencies to create their own characteristic sound. And all of these are overlapping yet still distinguishable – that is what we mean by a high bandwidth.

So on Earth sound appears to be an excellent way to communicate large amounts of information quickly over a large area. And yet, given that animals don't actually have languages, we can't help wondering whether the animals actually *use* all this information potential. Is there something special about human acoustic perception that animals don't possess? Something physically different about humans, so that we can harness the power of the acoustic modality for language, whereas birds, bats and dolphins can only tinker around the sidelines? We discuss this

★ Just search for 'online spectrogram', or download a free tool like the one provided by Cornell University called Raven Lite.

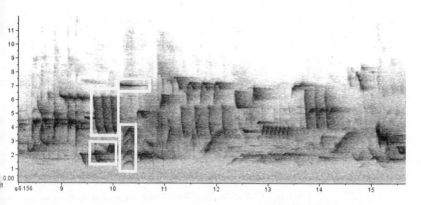

A spectrogram of the dawn chorus, with time from left to right, and pitch on the vertical axis. Many birds are singing overlapping songs, but they differ in pitch, so they can still be clearly distinguished (see boxes).

more in Chapter Nine, but interestingly the answer is probably not. Many other factors influence whether a particular species in a particular niche will evolve a language, but the fundamental physical mechanisms for producing and, importantly, interpreting sound seem to be shared across a wide range of animals. Certainly most vertebrates, and birds and mammals in particular, have a sophisticated mechanism in the ear – a 'frequency analyser' called the cochlea – that is capable of separating the different frequencies present in an acoustic signal, very much in the way that the image of the spectrogram illustrates.

Humans are not special in their ability to detect and discriminate between complex variations in sounds, and although our abilities are quite prodigious – consider how you can recognize the voice of your children or your friends across a crowded and loud room – they do not appear to be exceptional in the animal world. King penguins can recognize the calls of their chick among a colony of tens of thousands! Our ability to distinguish sounds of different pitches is particularly impressive when you consider that anyone can place the notes of a musical scale in ascending order of frequency, but no one can intuitively place the colours of the spectrum in ascending frequency.★ This again is the result of some universal constraints from the laws of physics. Sound waves can move tiny hairs on the cochlea, and different frequencies of sound move different hairs, which give us our perception of different pitches. But just as light has too short a wavelength to manoeuvre around bushes, similarly it usually can't make hairs – or anything much bigger than an atom – move. So, as long as signals do not have to traverse a vacuum, which sound cannot do, the acoustic modality seems to be a very promising channel for any kind of communication, irrespective of the particular evolutionary history of a planet. When we finally do discover aliens, we should not be surprised if they 'speak' with sound, even if everything else about them is, frankly, alien.

Nonetheless, we are faced with a peculiar dilemma. We have seen the decided advantages of the acoustic modality for communication. Highly intelligent dolphins communicate with sound. Cooperative wolves communicate with sound. Songbirds are vocal virtuosos. But compared to these animals, our own nearest relatives, the great apes, are vocal rookies. How did we become so vocal, when gorillas are effectively mute? Scientists have tried to teach chimpanzees and bonobos to make the sounds of human speech, but this seems to be beyond their vocal capacity. Is it possible that the ancestors we share with great apes used some primitive acoustic signalling, which evolved into our own language, but disappeared over the generations in the lineages of our primate cousins? We share quite a recent common ancestor with modern chimpanzees; that ancestor lived about 6 million years ago. Both we and modern chimpanzees share an ancestor with the modern gorilla, but this more distant grandparent lived

★ This was noted by the celebrated biologist J. B. S. Haldane in his 1927 essay 'Possible Worlds'.

something like 10 million years ago. However, our common ancestor with wolves and dolphins lived vastly before that, perhaps 95 million years ago – certainly while dinosaurs still walked the Earth – and our evolutionary history diverged from birds 320 million years ago, not long after our ancestors crawled out of the ocean. So we can hardly claim that the acoustic basis of our language skill is something that we inherited together with the vocal abilities of wolves and dolphins, let alone birds. Rather, vocal communication seems to have evolved rapidly in our chimp-like ancestors, diverging from their mute tree-dwelling relatives, and converging on the familiar vocal abilities we see in other, less related, animals. The convergence of vocal abilities is widespread on Earth, and other planets are likely to show similar evolutionary trajectories.

Light: communicating by seeing

Far more likely is that our ancestors 6 million years ago communicated primarily using visual signs and signals.* Studies on captive great apes have shown their particular aptitude for learning various forms of sign language, and even if there is considerable debate over whether these are true *languages*, there seems no doubt that our closest relatives are primarily visual communicators. Chimpanzees can be taught quite a complex series of hand gestures representing a wide range of concepts, and this ability seems to be more meaningful than just rote learning. Chimps even develop the peculiar habit of talking to themselves in hand gestures, if no other animals are around. So why *don't* we see any fully fledged visual language (apart from those recently invented by our own exceptional species)? Why did our ancestors

* See *The Evolution of Language* by W. Tecumseh Fitch.

appear to abandon their visual communication in favour of the acoustic medium?

Visual signalling is extremely widespread across the animal world: male birds flash their coloured feathers at potential mates; butterflies have large 'eyespots' on their wings to scare predators; male mandrills have absurdly blue and red snouts; and many animals like skunks and ladybirds use striking patterns to warn enemies of the dangers they pose. The bee waggle dance is even more sophisticated, giving information to their hive-mates on the location of food. The prevalence of visual signalling probably arises from vision itself, from the fact that it is incredibly useful to be able to see in the first place. Light-sensing bacteria are surprisingly common, and incredibly ancient – too ancient for us to be able to put a reliable age on the origin of a light-sensing ability. But from studying the proteins used to detect light in modern animal eyes, we do know that animal vision arose not long after animals themselves, about 700 million years ago. Even the common ancestor we share with jellyfish probably could see. So there has been plenty of time for vision to be adapted to a whole host of signalling strategies.

Light has many advantages for a signalling modality. Like sound, it is fast (even faster, of course, but there are hardly any cases where the difference would be relevant, even on other planets). Eyes evolved for many purposes, the most important of which were finding food and avoiding becoming someone else's food, so a light-detection mechanism was already well established when visual signalling became a possibility. Light also comes in different colours (frequencies), and that can give some added information to a visual signal, in a similar way to the combination of frequencies in an acoustic channel. Finally, light travels in very straight lines. What this means is that your light-sensing system (in our case, our eyes), can distinguish between light sources that are only very slightly separated in space, say, the tip of your thumb and the base. This provides an extra layer of information content to a visual signal: you can tell the difference between a thumbs-up and a thumbs-down. Sound, in contrast, tends to spread out like ripples in all directions, and making a spatial map of sound sources is extremely difficult. Even those animals that can detect sound sources with great accuracy, like owls finding prey in the undergrowth, or arctic foxes diving into snowdrifts to catch a rodent, still have only the most rudimentary ability to distinguish signal sources compared to your perception of the subtle difference between the letters 'e' and 'c' on this page.

But there the advantages of visual communication pretty much end, and the disadvantages begin. Light travels in straight lines because of its short wavelength compared to sound, but as we've seen, that same short wavelength means it is blocked by just about any obstacle. We cannot see through walls, trees, soil or clouds, and neither can animals, no matter on what planet they live. Light is great when you are next to someone and communicating directly with them, but as you become more distant, chances are something will be in the way, and visual communication then becomes impossible. Light is also strongly scattered even in transparent media like air, and in water particularly so. Even in the most pristine, crystal-clear seas, variations in salinity, water currents and even microscopic animals suspended in the water limit visibility. Typically, nothing can be seen at a distance of more than a handful of metres. Although the atmosphere of our planet is relatively transparent, other planetary bodies are less lucky; the low temperatures of Jupiter and Saturn mean that clouds of crystals of ammonia and other chemicals probably make the atmosphere opaque at most altitudes. We have no good idea what atmospheres are like on other exoplanets, but we can't assume they'll be as beautifully clear as our own.

Even if you are lucky enough to live in a perfectly clear

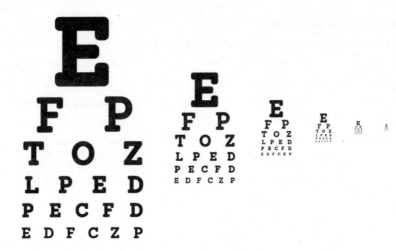

Although all signals become weaker as they get farther away, these progressively smaller images show how there are geometric constraints on visual information, in addition to any difficulty of visibility. Just being farther away means that signals will be less easy to interpret.

atmosphere with no trees or other obstacles, light communication suffers a further disadvantage. That same information-containing difference between the top and bottom of your thumb becomes harder and harder to distinguish at longer ranges – not only does the light get scattered, but there are also geometric considerations. You cannot read the writing on this book from a distance of more than a couple of metres; it's too small for the resolution of your vision system. Even if you can see the light, it is hard to get the information out of it, and this would appear to be a constraint that would apply equally on this world and in any alien environment.

What kind of workarounds could alien (or Earthling) animals use to offset the disadvantages of visual communication? Of course, visual signalling may be restricted to animals that are close to each other, where there are unlikely to be obstacles and the spatial discrimination is very clear, and this is exactly what happens in most species on Earth that predominantly communicate through light, including our nearest relatives. Some animals, such as fireflies, produce their own light, and very intense light at that. This bright signal can be seen from far away and is quite sufficient to do its job: to attract a mate.

But only the simplest of signals can be transmitted in this way. By flashing the light in particular patterns a degree of extra information can be embedded in the signal, in a similar way to Morse code, or the drumming behaviour of wolf spiders and kangaroo rats that use different patterns of on/off signals to identify which species they are, or even which individual. But the complexity of the information an animal can include in such a signal is limited. If a single 'bit', a dash or a dot, is missed from a Morse code signal, the message is corrupted. Therefore, pulsatile signallers keep things simple, for example by using the speed of the light flashes to indicate the attractiveness of the male doing the signalling. Light seems a very restricted modality for complex communication.

One group of animals worth mentioning in this context, however, are the cephalopods: octopuses, squids and cuttlefish. Many cephalopods have specialized skin cells called chromatophores that can actively change colour under the control of the animal's advanced nervous system. Interestingly, not all of this control comes from the brain, although cephalopods possess by far the most sophisticated brains of any invertebrates, but some of the colour-changing responses appear to be equivalent to what we would term reflexes. In any case, the dizzying range of patterns that cuttlefish in particular use is of such complexity and fast-changing dynamics that it is hard to avoid the thought that these colours form the basis of complex language. In fact, we are pretty sure that these psychedelic colour-changing displays are not language per se, but are used to convey fairly basic emotional states to other cuttlefish and, bizarrely, to confuse and essentially hypnotize prey before pouncing on them.

However, the fact of whether or not cuttlefish have a colour-language is not the point. The complexity of the visual signals in the swirling, pulsating colour patterns on the canvas of cuttlefish skin indicate quite clearly that very large amounts of information *could* be contained in such communication. If the other disadvantages of visual communication could be bypassed on some alien world (perhaps there are no trees or other obstacles to block the signals there), there is no doubt that a visual language based on complex coloured patterns is an evolutionary possibility.

But all this discussion of complex visual signalling and cuttlefish with hypnotizing colour swirls doesn't touch on one of the more ubiquitous but less conscious forms of visual signalling: what we loosely refer to as 'body language'. We use body language every day with other humans and are barely aware of doing so. Countless self-help books will tell you how to manipulate your body language to project confidence, dominance or attractiveness to potential mates, so there does appear to be a certain degree of conscious control over how we send visual signals to others.

We can appreciate the power of these kinds of subtle visual cues by thinking carefully about our communication with our pets. Dogs, in particular, have evolved for tens of thousands of years alongside humans, and both species have become particularly attuned to the unspoken communication of the other. If you have a dog of your own, you are inevitably conscious of his or her feelings: happiness, sadness, excitement, hunger, frustration. Apart from certain vocal interactions, this is all conveyed through body language: the position and movement of the tail, orientation of the ears, and that big sad look in their eyes when they want something.

But notice that this incredibly effective and complex communication channel breaks down somewhat when you are interacting with an unfamiliar dog, particularly if you have no experience of dogs of your own. Unless you are an especially keen observer of dog behaviour, then when you meet an unfamiliar animal in the street you are not guaranteed to understand what they think of you, nor are they likely to be able to interpret your intentions towards them. Our bond with our own personal companion is precisely that: personal. Body language can only be generalized up to a point, before individual personalities and idiosyncrasies swamp the meaning inherent in the signal. Far from having an agreed lexicon, visual signals can contain too much information, too much variation between messages with identical meanings. Although the visual modality itself is not inherently problematic for carrying a true language, perhaps under the conditions on our planet, and with the kind of animals that have evolved here, it has just been too unstable and too fragile to develop further. On other planets, a lack of covering vegetation plus a thin atmosphere (like on Mars) that both makes sound less effective and could also produce crystal-clear skies might be more conducive to animals with complex visual languages.

Smells: the oldest modality

Both sound and sight are deeply familiar to us as humans, and it doesn't take much to imagine an alien-inhabited world full of vocal and visual communicators. But neither sound nor light is the oldest signalling modality on Earth. The original and most ancient communication channel is one that we find very difficult to imagine developing into a language; in fact, we often fail to notice it completely. That modality is smell. Animals smell – a lot. Even bacteria 'smell', if we widen the definition to its natural limits, that of sensing the chemicals in the environment around us. The very earliest life forms would have gained a huge advantage from being able to follow the concentration of food chemicals in the water around them and so, rather than blundering around blindly, evolved to 'follow their nose' (even though they didn't yet have actual noses).

As with vision, once organisms develop mechanisms for sensing something important in the environment (light, food), then that mechanism can be co-opted for signalling, and this is precisely what happened, very early on indeed in the history of life on Earth. Even the interaction between different cells in an individual's body is mediated by chemical signals, and so 'chemical communication' in the broadest sense dates back at least to the origin of multicellular life, possibly as long ago as 3.5 billion years. Today, chemical signalling is close to ubiquitous across all animal life. So why is there no chemical language, in the sense of a true language? Why can you not write a poem in smells? And is this stunning lack of sophisticated chemical communication merely a fluke of Earth's environmental and developmental history, or can we expect that every planet we visit will be similarly devoid of flatulent Shakespeares?

The idea of a smell-based language may sound ridiculous because you might think that there simply are not enough distinct smells – chemical compounds – to supply the huge variety of concepts that we use in our own language – words, essentially. However, this may not be true. Even with a modest number of distinct smells, the number of possible combinations is huge. We know that our own rather unimpressive noses have detectors for about 400 different chemicals, dogs have 800 and rats can detect as many as 1,200 distinct stimuli. That means we have the ability – in theory – to detect about 10^{120} different chemical combinations – many, many more than the number of atoms in the entire universe.* Although this does not necessarily mean that

---

* With 400 different smell detectors, each of which could be 'on' or 'off', there are 2^{400} different combinations, or about 2.5×10^{120}. There are only about 10^{82} atoms in the observable universe.

we can consciously distinguish between any and all of those possible combinations of chemicals, at the very least we can say that a chemical modality could theoretically have the necessary complexity to transfer information on a scale we associate with language.

In much the same way as the cochlea breaks down sound into component frequencies, the nose's 400 separate detectors send separate messages to the olfactory bulb in the brain, which integrates these into a perceived 'smell'. Just by looking at the parallels between the action of the cochlea and that of the olfactory bulb, there is no neurological reason to think that a smell-language should be impossible. Insects are, of course, the Earth's champions of complex chemical communication. Smells are used to attract mates, to identify members of one's own colony, to mark the path to food, and to signal an alarm in case of an intruder. In many cases, even when a relatively small number of active chemical compounds have been identified, perhaps twenty, we can see that closely related insect species combine those compounds slightly differently, so that the messages of one species aren't confused with those of another.

However, as with our other modalities, the chemical sense must meet certain physical conditions if it is to be a candidate for complex communication. Sight and sound are fast – chemical signals are not. A firefly's flash reaches its recipient instantaneously; a cricket's chirp perhaps with a delay of a second or two. At any scale larger than that of a few centimetres, the speed at which chemicals spread out from their source is hundreds, if not thousands of times slower. Although it is almost impossible to quantify the 'speed of smell', it is usually true that passive diffusion is much slower than a smell borne on the wind. So, one might consider the absolute upper limit to the speed of smell to be the speed of the wind: typically of the order of 10m/s compared to sound at 340m/s. If you are waiting for your wind-borne message to arrive from a signaller on the other side of the road, you may be in luck on a blustery day (at Beaufort 6, 'strong breeze', wind speed 13m/s) it could take a second or two, but on a still summer evening (Beaufort 0, 'calm', wind speed 0.3m/s), you could be waiting a minute or more to get the message. Of course, on a planet where winds are regularly strong and reliable, perhaps chemical signalling could provide a fast communication channel. Unfortunately, it would be an exceptionally one-way channel – good luck getting your reply back to the sender when your smells are fighting upstream against a gale!

The simplistic idea of speeding up chemical communication using wind brings other problems. When air is flowing slowly over a smooth surface, it tends to flow in a straight line, and smells can be carried directly from their source to the animal sensing them. But as the wind speed picks up, or if the surface is rougher, the air tends to break up into tiny whirlpools or vortices and eventually degenerates into a messy squall of wind currents blowing in all kinds of different directions. Any delicate combination of different smells, perhaps carefully deposited at different locations to give subtle complexity to the signal, would be utterly blended and homogenized like a single drop of food colouring in some cake batter. Keeping those chemical signals apart while dispersing them across the environment may be a crucial limitation in the evolution of a chemical language. Any alien animals building their technological civilization on science textbooks written in smells may simply be constrained to be communicating over the short distances where smells don't get mixed up. They would be very small aliens indeed.

Electricity: the language of life

We must now leave behind the comfortable and familiar worlds of sight, sound and smell, and venture into a modality so 'alien' to us that it is difficult for us to comprehend how animals that live with this almost science fiction-like sense even perceive the world. Foremost among these are the incredible electric fishes of Africa and South America. If any Earthling animal species gives us a radically different insight into potential alien communication systems, one not influenced by our own preconceptions of what perception is, it must surely be these.

Electricity is absolutely fundamental to life on Earth. All life needs to store energy and move energy around an organism. On Earth, without exception, this is done by moving electric charge, positive and negative, within and between cells. Charges exert forces on each other (like charges repel, opposite charges attract), so energy is needed to move charge through an electric field, just like moving mass through a gravitational field – think of pushing a car up a hill. One could speculate wildly that alien planets may evolve life that makes use of some other field to store energy, perhaps even gravitational fields, but with our current knowledge of physics, none of these seem very likely: gravity really isn't very powerful, except for very large objects.

Anyway, on Earth, life means electricity. If all life generates electricity, then sooner or later some life form will evolve the ability to detect electricity so that they can hunt down other organisms and gobble them up. Electroreception, or the ability to detect electric fields, is widespread among many different species of fish, including sharks, but is also found in amphibians, like salamanders, and, bizarrely, in some mammals, like duck-billed platypuses, which hunt out their prey by detecting the electric signs of life penetrating through their muddy environment.

Given that the water is filled with electricity-hunting predators, it would seem a reckless move to send out intentional electrical signals as a form of communication. But despite that, some species of fish, particularly the South American knifefish and the African elephantfish, have evolved specialized electric organs within their bodies that generate complex varying electric fields in the water around them. Rather like a stack of button batteries, specialized cells called electrocytes each generate a small voltage, but by connecting them one after the other, quite strong signals can be produced. As these cells are derived from muscle cells (which have among the most powerful electrical activity in the body), the fish have very fine control over their electrical activity, just as you have fine control over your speech by using the muscles in your larynx and tongue. Naturally, these fish, like sharks and their other enemies, have also evolved electroreceptors: specialized cells that detect the electric field around them.

Electric fishes use their generated electric fields for two things. Firstly, they can sense the environment around them by detecting disturbances in the electric field caused by inanimate as well as animate objects. When they approach a rock, for example, the pattern of the electric field around them is altered slightly, and the brain perceives this in much the same way that we perceive objects by sensing light. How peculiar it must be to see the world through the distortion of electric fields! This is a sensory experience we cannot directly understand, although perhaps it would be possible to design a simulator that converts electric-field distortions into visual signals for our amusement.

However, of particular interest to us is the way that electric fishes use electric fields for communication. Although the African and American species encode information into their electrical signals in somewhat different ways, the basic principle is the same: the electric field is intentionally varied, or

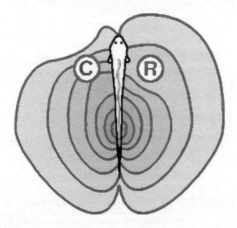

Active electrosensing in fishes. The electromagnetic field (shown by the lines) is distorted differently around different types of objects: those that conduct electricity (C) and those that resist electric flow (R). The fish translates these distortions into a map of the world around it.

'modulated', to give a characteristic pattern. Subtle differences in pulse or wave patterns can distinguish between different species of fish, between male and female, and can even indicate the social status of an animal and its dominance position. In the dark, murky and muddy waters of jungle rivers, vision is nearly useless, and this complex electrical mechanism for both sensing the world around you and communicating with others of your species seems ideal.

Certainly, the potential complexity of electrical signals is comfortably high enough to support a complex language. Electric fields travel fast and the signaller can be localized easily – not surprisingly, as we know fish also use their fields to navigate around passive objects. Finally, the waveform could (in theory, although it is not clear that fish do this) be modulated with different wavelengths to ensure that signals are not blocked by typical obstacles in the environment, so that communication could be effective

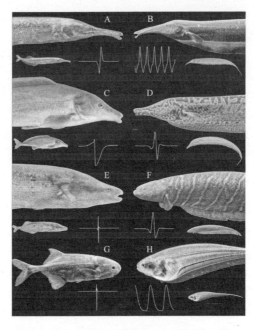

The different modulation patterns of electric pulses in different species of electric fishes. At the very least, the complex patterns give information on the species type, but in principle far more information could potentially be encoded in such a signal.

over reasonable ranges: more than the handful of metres over which electric fishes communicate.

All in all, electric communication seems an almost ideal modality for the evolution of a language. Yet we are certain that no electric fishes have a language, and more worryingly, these two families of fishes are the only ones to have a complex electrical signalling system at all. Why is electrical communication not more widespread and more sophisticated in the animal world? Could the barriers to wider adoption of this telepathy-like communication on Earth mean that aliens too are unlikely to speak in electricity?

There seem to be two reasons why we see so few creatures on

Earth with a system of electric signals as sophisticated as electric fishes. Firstly, it is a very expensive system to evolve and to maintain. Generating powerful electrical signals requires a lot of energy, and a large proportion of an animal's brain must also be dedicated to decoding and interpreting the complex signals from its array of electrical detectors. In short, electrical communication is only likely to evolve if there is very powerful evolutionary pressure; essentially, if the animals have little choice. In a similar way, bats evolved echolocation – another really useful trick – because there was no other way they could take advantage of the cave-dwelling and night-feeding niche that they exploit. So electric fishes are like the specialist you pay to fix your car. They have the specialist tools needed to solve particular problems; tools that you might find useful if you had them, but it's not worth going out to buy them yourself.

The second reason why electric communication is not more widespread is due to the constraints of physics. Electric fields exist in materials that are not too conductive (so metal wouldn't work well) and not too insulating (air would be a poor candidate too). Although we do not have many metal environments on this planet, we do have a lot of air, and animals that live out of water would be unable to maintain the kind of static electrical fields that electric fishes use to detect the objects around them. As we have seen with each of the other modalities, evolution builds complex solutions on top of simple ones, taking sensory systems like vision, hearing and smell, and adapting them for communication. If electric sensing does not work on land, electric communication will not evolve there. Interestingly, those land animals that *do* use electric sensing, like the echidna, have evolved very particular solutions to this problem, such as coating their electric-sensitive snout with copious amounts of conductive mucus, but these examples are perhaps too few and far between to lead to communicative innovation. So, on our planet, electric

The short-beaked echidna: an egg-laying mammal whose mucus-covered snout is used for detecting the electrical signals of its ant and termite prey.

communication is either impractical (on land) or unnecessary (where water is clear enough to use other senses). The kind of planet where one might expect electric communication to evolve would be one where the oceans are completely dark; and as at least two such worlds are thought to exist in our own solar system (Titan and Enceladus, the moons of Saturn) such environments may in fact be quite common.

We cannot be sure that we have considered all the possible communication modalities that could be candidates for carrying complex information. Some, like using magnets to send messages, we simply do not observe animals using on Earth, and so we have little to say about what may or may not drive their evolution. Others have been postulated by philosophers or science fiction writers and appear quite plausible on paper. In Fred Hoyle's novel *The Black Cloud*, currents of ionized gas were used to send messages between the distributed 'organs' of an interstellar creature millions of kilometres across. Speculation – or at least, fruitful speculation – is not exactly easy, but it carries with it little burden of validation.

What we have seen in this chapter is that a wealth of evolutionary diversity on our own planet has exploited not just those modalities that seem to us the obvious candidates for language, but almost every communication strategy that the environment on Earth supports. Even if an alien planet differs drastically from Earth in its atmosphere, its temperature and pressure, even in the very elements and molecules that make up its surface, at least we can draw on our Earthly observations of how animals adapt to draw conclusions about what aliens might – or might not – use to talk to one another.

On a dark, subterranean world, perhaps like the underground oceans of Enceladus, vision may be totally absent, and eyeless creatures could evolve a perfectly competent and rich communication using sound alone. Conversely, in the tenuous Martian atmosphere, acoustic communication just isn't a good option.

As we saw in the previous chapter, Earth is an evolutionary testing ground for the fairly limited number of realistic solutions to life's problems that the laws of physics can provide. Aliens in the dark will click like bats and dolphins, and aliens in the clear skies will flash their colours at each other. If an alien world is broadly similar to ours physically, we would likely experience a similar rush of stimuli in different modalities as we do when taking a walk in the woods of Earth.

6.   Intelligence (Whatever That Is)

My dog sometimes winks at me. He's just sitting there looking at me calmly, and then he does it. He winks. I look around surreptitiously in case anyone is watching. And I wink back. What can I do? If he's just an automaton – a robot following instinctive rules – what do I have to lose? But what if not? What if he is a conscious, intelligent being that knows perfectly well what *I* mean by winking, and is trying to say, '*I'm more intelligent than they think, and I just want to check that you know that. Wink back to let me know you're in on the secret.*'

Most of us don't really believe that our animal acquaintances are capable of such thoughts, and many would go so far as to say that animal intelligence – if they even have any – doesn't rival ours. Surely our achievements, both personal and those of our civilization, and the animals' lack of similar achievements, are evidence that there is a qualitative difference between our cognitive abilities? We are smart. They are . . . less smart?

We have been fascinated for millennia by animal intelligence, and how it differs from human intelligence.* But what, if anything, is *fundamental* to intelligence? What features would we expect to see in any and every organism on Earth and on other planets that make us say, 'Yes: that creature is intelligent'? These features could be particular behaviours, or particular abilities. Or maybe we should define intelligence in terms of particular

* Books by Frans de Waal, Gregory Berns and Justin Gregg, given in the Further Reading section of this book, are all highly entertaining and thought-provoking views on the question of animal intelligence.

kinds of brains, or the way in which such brains would be programmed. However, it should be no surprise by now that I am going to claim that the most interesting universal feature of intelligence is how it *evolved*. After all, it is shared evolutionary mechanisms that are going to determine whether – or not – we can find common ground with our alien neighbours when we meet them.

Finding an acceptable definition of intelligence has a chequered history, and some attempts to measure intelligence in a quantitative way are downright suspect. Humans have used supposedly objective intelligence measures to claim their superiority over other people (and in particular in denouncing races that they don't like), and also to propose that there is a uniqueness to humans that distinguishes them from other animals. But we are not so much looking for a definition of intelligence; rather a general framework of how intelligence might evolve on another planet, and whether the path that such evolution would take is similar to that on Earth. Will technological aliens have an intelligence that we would recognize? Or are there many different ways of being intelligent and at the same time building a radio telescope? Are we looking for 'human-like' intelligence, and is that perhaps precisely what we mean by intelligence? Justin Gregg writes in his book *Are Dolphins Really Smart?* that the definition of intelligence that best fits our intuitive understanding that flies are stupid and chimpanzees are smart is that 'Intelligence is a measure of how closely a thing's behaviour resembles the behaviour of an adult human.' Or is a human-centric definition liable to lead us down dark alleyways where our future alien interlocutors are ignored, and possibly even overlooked?

Human intelligence does appear to be unusual on this world, and there are fascinating questions that still have not been answered about what makes an Einstein or a Mozart. But we need to strip away a lot of these details about humans and their particular intelligence and go back to evolutionary basics. We want to know what Einstein, Mozart and every one of us have in common, and how that evolved from our ancestors who didn't possess 'that', but possessed some other kind of intelligence – about which we also want to know. To put human intelligence in its proper context we must attend to the millions of other species of animal on Earth, each of which possesses intelligence of their own. It would be woefully closed-minded to think that evolution has been working hard for 3.5 billion years simply to produce this long-awaited fruit: human intelligence. So can we provide a way of looking at intelligence that simultaneously applies to all creatures, from sponges to humans, capturing a process that is general enough to apply on other planets as well? If we cannot compare between species on this planet, then we will not have identified the truly universal features of intelligence that we will need to use to predict intelligence in life on other planets.

But even before we consider what intelligence is, we should pause to think about *why* intelligence is. Intelligence is fundamentally about solving problems, and this makes sense, because the ability to solve the problems an animal faces would seem to be something that evolution could favour. The world – every world – is full of problems. Energy is limited; space is limited; time is limited. Figuring out how to exploit those limited resources is a problem, and the ability to overcome these problems could provide one individual with an advantage over others.

For example, the amoeba moves towards nutrients in its environment by following the direction of maximum concentration: slithering left if food is more concentrated to the left, or right if there's more food smell on the right. For many, this kind of simplistic behaviour would stretch the definition of intelligence somewhat. Of course, the amoebae don't 'think' in any way, having no mechanism like a brain to do the thinking. However, there is an undeniable subtlety to the behaviour of amoebae.

A creature even simpler than an amoeba might follow an even simpler rule: 'Move ahead, eating whatever is in your path – don't move left or right.' But the amoeba has improved on this; 'ahead' may not be the best way to go. Food may be behind you. The world is, after all, *unpredictable*. Our single-celled organism waits until it senses the concentration of nutrients before deciding which way to move. This intelligence, as well as most intelligent behaviours in animals (and also in humans) can be explained without recourse to 'thoughts' or 'mental processes'.

Predicting the nature of the world does seem to be a big part of the way that animals use intelligence. Will that lion attack me, or is it too far away? Should I fly south for the winter now that the days are getting shorter? I wonder if that male is going to make a good father to my offspring? It seems inevitable that *every* kind of intelligence must be able to iron out some of the unpredictability of the universe. We humans have taken this to extreme lengths: will this space probe enter orbit around Jupiter? Will the universe continue to expand forever, or will it collapse? Can I threaten to use nuclear weapons without my bluff being called? One popular definition of human intelligence is that we have the ability to build models of the universe in our heads and to predict what will happen under different possible scenarios.★ We can 'run mental simulations', testing out possible solutions to real-world problems, but without any of the risks associated with trying them out in real life. Certainly we have a very special ability to do this, and it is an ability that most other animals probably lack, but do they lack this ability in degree or in kind? Is our intelligence just better, or is it fundamentally different?

Let us begin with the observation that our intelligence, and

★ The philosopher Daniel Dennett builds a very clear and useful hierarchy of different types of predictive powers in his book *Kinds of Minds: The Origins of Consciousness*.

that of other animals, is a kind of prediction machine: brains are predictors. So the precise evolutionary trajectory that an intelligence will take is deeply dependent on exactly what it is that the creature is trying to predict. We may be interested in whether our prey will escape us, or whether we will escape our predators, and so we live in a world of moving visual objects, and our brains have created from that an internal representation of how objects move in the world. An electric fish, such as we discussed in the last chapter, lives in a world utterly dissimilar to ours and possesses a completely different intelligence about how to interpret and predict objects. Subtle variations in the electric field around its body may be interpreted as the positions and movements of objects, but if we were to understand how an electric fish perceives the world (and we do not) then it seems likely that their very concept of 'spatial relationships' between objects would be completely different from our own. How much more so for aliens inhabiting very alien sensory environments.

In a famous essay, the philosopher Thomas Nagel proposed that it is not enough to try to imagine how a bat perceives the world.★ At the very best, we could understand only a human's translation of a bat's perceptual experiences. The 'batness' of the experience would be completely beyond us.

This is disturbing, because it suggests that there may be a myriad diversity of types of intelligence, each one so fundamentally different from the other that there is, to all intents and purposes, nothing in common between them. If we cannot even perceive in what way a bat is intelligent, then surely there is no hope of putting ourselves in the mind of an alien. On a planet where life evolved in the darkness of an underground ocean, the intelligent creatures there surely have concepts and comprehensions that are

★ See 'What is it like to be a bat?' by Thomas Nagel.

utterly unimaginable to us, just as our ideas of sunsets and rainbows would be impossible to convey to them.

Fortunately, the study of animal behaviour over the past fifty years has uncovered some vital clues that help us to understand where the commonality lies between our intelligence and that of all other creatures on the planet. Intelligence in animals seems to involve *integrating* different sensory perceptions and different predictive skills, to make sense of and to anticipate the outside world. But how is this done?

Many intelligences, or just one intelligence?

Much of our uncertainty over what alien intelligence will be like arises from our uncertainty about the nature of intelligence on Earth. With regards to this, scientists have suggested at least two possibilities. Is intelligence a *general* capability, something that can equally well be put to use for wildly different challenges; for instance, both to solve equations and to catch a flying tennis ball? Or is intelligence made up of multiple different *specific* skills; one for solving equations, and one for catching balls? If intelligence is general, then we can all perform all tasks requiring intelligence, but some people will be better than others. Alternatively, if intelligence is specific, any particular type of intelligence could be completely absent in some species, and even in some individuals. What does it tell us that some animals (and humans) can catch balls with precision, but are hopelessly bamboozled by mathematics, and vice versa for others? For the record, my dog can do neither. His genius clearly lies in some as yet undiscovered field.

This question is crucial in understanding the nature of alien intelligence. If intelligence is a general (and universal) ability, varying only in degree, rather than in kind, we would expect intelligent aliens to be intelligent in much the same way as us.

However, if intelligence is based on specific abilities, and is linked to the specific problems an animal needs to solve – such as an archerfish accurately spitting out a jet of water to shoot down an insect perched above – then it is quite possible that alien intelligence will be based on experiences so fundamentally different from ours that mutual understanding between us, or even recognizing the presence of intelligence in the other, may, unfortunately, be impossible.

Scientists have pondered this dilemma for many years: do animals evolve a range of very specific intelligent abilities to solve specific problems in their environment, or is intelligence one general property that is widely put to use, including in solving those same specific problems? Like all dichotomies, this is almost certainly a false dichotomy, but nonetheless, the arguments for and against each have revealed a great deal about the different ways that evolution may and may not operate towards the development of intelligence.

It may seem obvious that different species have evolved different capabilities: living in water or on land, eating plants or chasing other animals, and so on. As the problems that needed to be solved in each of these very different niches are in themselves very different, those animals possess different 'intelligences' for solving their particular problems. This seems uncontroversial on the face of it: the archerfish can accurately hit an insect, despite the refraction of light through the surface of the water; a bat can catch flying insects using a phenomenal capability to predict their movements in three dimensions; and a colony of leafcutter ants can construct a nest many metres deep that contains specialist chambers where the ants farm a specially domesticated fungus. Surely these are examples of different intelligences, and could not be considered variations on a single theme?

However, this position, taken to its logical conclusion, would argue that humans are not, in fact, *more* intelligent than jellyfish;

we are just intelligent 'in different ways'. This claim would be counterintuitive to most people; at the very least, we can say that humans appear to be intelligent in multiple different ways: we use logic to build scientific understanding of the world around us, create unique compositions of art and music, and use our social abilities to flourish in cities and countries consisting of millions of individuals. Jellyfish, on the other hand, barely cut it in the 'floating through the sea' category. But claiming that humans 'must' be more intelligent than jellyfish is a somewhat circular argument: if we assume from the start that intelligence is 'what we (humans) have', then inevitably we will have more than other animals. A more objective approach is to look at what kind of intelligence we would expect to evolve and why, and then reassess how that property is distributed throughout our ecosystem.

The argument that there is a single kind of intelligence that varies between species primarily in degree, rather than in type, rests largely on an observation from the study of human psychology, something that has been somewhat tentatively observed in other animals too. Psychologists suggest that many different measures of intelligence appear to be correlated with each other. People who are good at maths tend to also be good at languages and music. This (they say) implies that there is something in common going on in the brains of these 'clever' people. But many of the original studies into 'general intelligence', or 'g-factor' (dating back to the 1900s) have, since the 1980s, been largely debunked.★ Testing human intelligence in a way that controls for upbringing, socio-economic status, and even the cultural biases of the tester is an area hugely fraught with difficulties. At best, human intelligence testing has been used naively, assuming a scientific objectivity that probably does not exist, and at worst it has been used maliciously to promote division and racism. The

★ See *The Mismeasure of Man* by Stephen J. Gould.

idea that it is possible to achieve a single number like 'IQ' (an Intelligence Quotient) that summarizes an individual's intelligence is controversial. Does it make sense to extend the idea of IQ tests to animals, which have evolved to solve different kinds of problems, and use different kinds of information to humans? The idea seems farcical. Anyone who has seen videos of an octopus unscrewing the cap of a jar from within, has a feeling that this animal cannot be anything other than intelligent, but no IQ test would capture the nature and extent of that intelligence, and how similar it is to human intelligence, or how different.

The proponents of general intelligence do make one important point related to the evolution of intelligence, which we have to take seriously. Much intelligent behaviour depends on a few core abilities, specifically: learning, memory and the ability to make decisions. These abilities are ostensibly rather simple but seem to be necessary for many of the intelligent tasks that animals, including us, perform. We know that it is possible to train a huge variety of species to 'learn' certain behaviours, from rats finding their way around a maze, to fish that can recognize faces. The same psychological techniques that you use to teach your dog to sit can be used to teach a chicken to skateboard.

Famously, Ivan Pavlov observed in 1897 that dogs could be conditioned to respond to a completely neutral signal – a ringing bell – as if it indicated that a reward was expected. The dogs salivated when they heard the bell, because they had come to associate one with the other. But a bell has nothing to do with food, so this cannot be an instinctive or evolutionarily innate response. And yet, what could be more evolutionarily useful? If you can predict the arrival of food using some signal that isn't the food itself, you are clearly ahead of the game and have an advantage over your competitors.

In the years immediately following Pavlov, scientists uncovered another uncanny behaviour that seems to be shared by almost

all animals: an animal can learn to change its behaviour to produce a favourable outcome — even if that behaviour itself is
neutral. No one is surprised that a dog can discover food by
searching for food – that's just evolutionary adaptation. But you
can also teach a dog to sit by rewarding her with food. The fact
that a dog can learn to get food by *sitting* is exceptional. Sitting
has nothing to do with searching for food – it is behaviourally
neutral. Clearly, the dog has associated its own response to the
command 'sit' with the arrival of food. It has learned, and it has
learned a way to get food by being *flexible*. And that kind of flexibility seems to be a key part of what it means to be intelligent.

Amazingly, the ability to learn in this way seems exceptionally widespread across animals on Earth. Mammals like dogs and
monkeys learn, as do birds that recognize predators by watching
the responses of other birds, and fish that identify good feeding
patches from the feeding success of other fish. Even insects learn,
despite the fact that insect brains are minuscule. Researchers
recently trained bees to 'play football' by manipulating a small
ball into a 'goal' to get a sugary reward. The ubiquity and diversity of learning among animals on Earth – as well as its obvious
evolutionary benefits – make it seem almost certain that the
ability to learn to associate certain actions with certain outcomes
is a universal feature of intelligence. If alien animals have evolved
intelligence to solve problems on alien planets, then they must
have evolved the ability to associate actions and outcomes. Associative learning, as it is called, must be universal.

'Learning' is an appealing criterion for intelligence. We do not
generally consider animals to be intelligent if they are following
purely instinctive drives. A frog that catches an insect with its
tongue is undoubtedly doing something 'clever', but on the other
hand, it is 'merely' performing instinctive behaviours. The flexibility involved in learning something new is clearly distinct from
this – and is something that intuitively feels like a prerequisite for

intelligence. Is this, then, a universal definition? Intelligence is learning, no more, for animals on Earth and for aliens as well?

Most evolutionary biologists would not accept that we have reached the end of this particular enquiry. Observing animals in the real world reveals a very different picture from those that came out of the sterile laboratories of Pavlov and his intellectual descendants in the first half of the twentieth century. The image of white-coated scientists with clipboards watching rats running around a maze may be a huge part of our cultural perception of science, but it portrays only a part of the picture. In the logical desire to drill down to the most basic of behaviours, and to remove all possible distracting and irrelevant information, scientists placed animals in highly controlled environments and presented them with very specific tasks. Scientific experiments are always simplified, but that can sometimes be misleading.

Animals did not evolve in a university laboratory, but in the outside world, where sensory stimuli wash over them with dizzying diversity and, crucially, with conflicting information. Rats in the real world never have to choose between two identical paths, left and right, and they certainly did not evolve to do so. The results that come from studying animal behaviour in the wild often clash with those from the psychology laboratories, and arguments about the validity of these experiments continue to this day.

A key piece of evidence in our search for what is universal about intelligence is that in the wild different animal species behave very differently to each other. The mental skills needed for a chimpanzee to survive in the forests of East Africa are quite different from those needed by a crow on the island of New Caledonia in the South Pacific. Both are intelligent – among the 'most' intelligent animals on the planet – but do they both have the *same* kind of intelligence? Studying animal behaviour in the wild raises two problems with the idea that intelligence is no more than a general learning capability.

Firstly, looking at the incredibly advanced and complex behaviours of animals like the chimpanzee and the New Caledonian crow casts doubt on the idea that all the intelligence in the animal world could result from some kind of shared ability. In 1960, Jane Goodall described how chimpanzees fashioned tools out of sticks, to extract termites from a nest.* With this discovery, she rocked the foundations of supposed human uniqueness – we used to think that we were the only species that used tools! And Goodall's discovery opened the floodgates. New Caledonian crows are rather unremarkable-looking birds that live on the French colony of New Caledonia in the South Pacific. They too make tools, taking twigs and using their beaks to get them into the right shape for extracting insects from holes in logs. Furthermore, these crows have recently been shown to have another, even more surprising skill. They can make tools, *to make other tools*. In these experiments (admittedly, in a laboratory setting) the birds used a short stick to retrieve a longer stick that was necessary to get at food. These crows can also choose between tools that are likely to work, and those that won't, and can even fashion tools into different shapes with their beaks depending on the nature of the problem facing them.† It isn't just that we aren't the only ones making tools, we aren't even the only ones using technology.

That this is intelligence, there can be no doubt. But consider that tool-using chimpanzees and crows last shared a common ancestor 320 million years ago, at a time when the land was dominated by giant fern forests and giant insects, with dragonfly-like creatures up to a metre across. All mammals, from shrews to whales, as well as all modern reptiles and birds, share this same common ancestor. If intelligence comes from a common ancestor,

* See *Through a Window: My Thirty Years with the Chimpanzees of Gombe* by Jane Goodall.
† See *The Genius of Birds* by Jennifer Ackerman.

it should be found in all (or at least most) of its descendants. Why aren't all birds, mammals and reptiles that intelligent? Is it likely that the intelligence of the chimpanzee and the New Caledonian crow is inherited from this ancient ancestor, and that intelligence just withered away in the vast majority of other families of animals, like lizards, turtles, canaries, opossums and wildebeest?

Of course not. The only sensible explanation is that both chimps and crows evolved their exceptional intelligence afresh, as did many other species, including dolphins that cooperate to herd fish into shallow water, capuchin monkeys that use rocks to crack nuts, and of course humans, with all of our problem-solving capabilities. Intelligence evolves all the time to fit specific needs – it is not merely an inherited trait from the dawn of time. The patterns that we see here – intelligence evolving again and again and again to solve different problems in different domains – are a compelling indication that alien animals too will evolve problem-solving intelligence, on different planets throughout the galaxy. Earth creatures are not so uniquely special and smart.

This convergent evolution of intelligence raises the possibility that different mechanisms may be operating in different species. Rather than some inherited general intelligence being moulded into the impressive feats of both chimps and crows, these two species may have evolved their specific abilities from quite separate behavioural ingredients, perhaps leveraging their different sensory perceptions of the world, or building on different brain structures that have nothing in common with each other. This appeals to evolutionary biologists, because 'general' abilities (like general intelligence) are more difficult to explain evolutionarily than specific abilities. Why would an animal evolve to be generally good at something, when it could evolve to be specifically even better? Specialists are better at what they do than generalists; that's why brain surgery is performed by a neurosurgeon, not a GP. So if crows need to extract insects from trees, they might evolve that very specific tool-using skill, and that would serve them very well indeed. For a crow to evolve the ability to adapt that insect-extracting tool to other situations – probably situations they will never encounter in their lifetimes – seems wasteful. If it doesn't produce an improvement in their survival, and the number and success of their offspring, there simply isn't a mechanism for that ability to evolve.

The second problem with the suggestion that general intelligence is the common currency of intelligence itself (here on Earth and across the universe) is that many of the behaviours we identify as intelligent in animals are quite clearly based on separate and specific brain mechanisms. Birds are very good at learning, it is true, birdsong being the most obvious example, but birds are capable of learning at least two very different types of skills, which require different kinds of intelligence.

Many species of songbirds have an innate ability to sing, but if they are raised without the opportunity to listen to the songs of others, they grow up unable to sing properly themselves. They must learn the songs, that much is clear. Some species are capable of the most amazing feats of song rendition, remembering hugely complex songs made up of hundreds of different notes. When I worked at the University of Tennessee, I couldn't walk to my department without having to stop and listen to the unparalleled performances of the northern mockingbird. This species mimics the songs of tens of different species of other birds, possibly as many as a hundred. A single male can sing a few bars of the song of an American robin, then switch to a blue jay, then a Carolina wren, then a nuthatch, and on and on and on. This is not an inherited behaviour – each individual male learns the songs of those birds he hears around him as he grows up. It is unquestionably an impressive feat of learning, and therefore intelligence.

However, many species of birds have memory and learning skills that are essentially *different* in nature. Many birds that live in environments with warm summers and cold winters gather food when it is available, and then hide or cache the food in different locations to provide a resource during the lean months. The black-capped chickadee in the Rocky Mountains in North America can hide literally thousands of pine kernels in different locations, and then recover them one by one weeks later. This is an amazing feat of memory, surely beyond any human capability. And from studying the tiny brains of these amazing creatures it is clear that, whatever is going on in these small minds, it is something *different* from song learning.

At the very least, the memory for learning the location of cached food takes place in a different part of the brain from song learning. Birdsong is centred on a small and specialist structure in the brain that is, not surprisingly, absent in other less musically gifted species – they don't need it if they don't sing. Conversely, the ability to remember the location of cached food could be important to anyone, and seems to be associated with a structure called the hippocampus, present in the brains of all vertebrates, including humans, and enlarged in those individuals particularly skilled at remembering their cache locations. Tantalizingly, it has been suggested that the human hippocampus is larger in London taxi drivers who have to remember their way around all the streets of the capital. Alien creatures with completely different brain structures (no hippocampus, most likely) will have specific learning abilities centred on quite specific alien brain structures, evolved especially for those purposes.

Now, unlike general intelligence, biologists find it very easy to explain the evolution of these very specific learning abilities: the chickadee that can remember the locations of more of its cached seeds is less likely to starve in the winter. Intelligence that serves a specific purpose is inevitable and clear-cut, and surely exists on any planet. So should we abandon any ideas of general intelligence, and say that each animal's intelligence is whatever is specific for its own species and its own niche? Perhaps the goal of defining the universal characteristics of intelligence is a mirage, and one best avoided?

The most satisfying narrative is that general intelligence evolves and serves to integrate different kinds of specific intelligence. Some animals behave in a human-like intelligent way when they, like us, take many different features of their environment and process them in different ways, leading to a single conclusion. Scientists are increasingly viewing intelligence as a hierarchical process, in which different abilities are integrated at different levels. Each one of those abilities – remembering where you hid your seeds, or how to sing a song – is in itself a response to evolutionary pressures. However, when the environment is particularly unpredictable, each of the different abilities can be leveraged together, synergistically, to produce a performance that is more than the sum of its parts.

I believe that this is the best way to approach predicting what kind of intelligence we will find on other planets. Alien animals will likely possess similar specific abilities to those on our planet, assuming that the environmental conditions on those planets have parallels to those on Earth. On a planet with seasons, the creatures will probably cache food. On a planet with hard to access food, they will use tools. But the aliens that we tend to call 'intelligent aliens', those with the technological abilities to contact us, will have evolved to integrate those specific abilities and apply them to new and unique scenarios. Creatures that can put together their caching skills and their tool-building skills – realizing that this provides them with abilities beyond merely hiding seeds or fishing out grubs – will be the ones that end up figuring out how to build a spaceship.

Aliens as scientists and mathematicians

We scientists tend to assume that aliens will also be scientists and mathematicians; indeed, more advanced and skilful than we are ourselves. Otherwise, how could they build a spaceship to visit us, or radio telescopes to send messages to us? Popular science fiction would seem to concur, although all too often those alien scientists are performing experiments on hapless humans rather than benevolently sharing their wealth of knowledge with us. But I know philosophers who believe that aliens will be philosophers. Do electricians and plumbers think that alien civilizations will be as reliant on electrical and plumbing skills as we are?

Science has a history of biasing its methods and its findings by the cultural and social background of the scientists themselves. But although aliens may or may not have indoor plumbing or central heating, the laws of science and mathematics are the same for them as they are for us. Surely this is a common point around which we and alien civilizations can agree? Both human and alien scientists will have made many of the same discoveries, and alien mathematicians will have derived the same mathematical theorems as human mathematicians have done on Earth. If so, surely we can use the most fundamental ideas of logic, mathematics and science to build a common communication channel between ourselves and alien species, even if we are different in every other way?

Certainly, such ideas have been proposed ever since scientists and philosophers began to give serious consideration to the possibility of alien life. In the 1980s, the astronomer Carl Sagan wrote eloquently about the ways in which alien civilizations could use mathematical principles to establish communication with us,[*] and he himself (together with his wife, Linda Sagan, and Frank

---

[*] *Contact* by Carl Sagan.

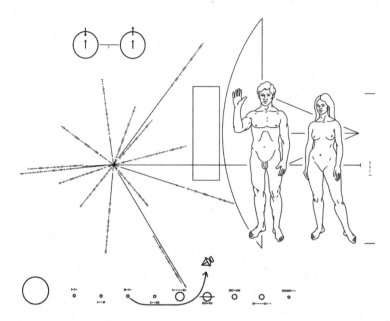

The Pioneer plaque, showing information about humans and our planet in a way that, hopefully, will be understandable to alien intelligence. This is assuming that the aliens understand the concept of 'numbers'.

Drake, the 'father' of the search for extraterrestrial intelligence) designed the famous Pioneer plaque that accompanied two tiny space probes launched in the early 1970s on their mission out of the solar system. As well as a visual representation of two human figures, the plaque gives mathematical representations of the unique rotation periods of fourteen prominent pulsar stars, as well as the directions from the Sun to each of those stars. Any civilization finding the plaque should be able to locate our solar system using this 'map'. So, perhaps mathematics can help us not only in the search for extraterrestrial intelligence, but also in designing messages to be broadcast into outer space to signal that we, too, are intelligent.

Since the 1960s, scientists have suggested that mathematics is a universal language, something inevitably shared between us and every alien civilization. The laws of mathematics are, after all, truly universal. If we try to communicate using these laws, then we are guaranteed at the very least not to be talking nonsense. A triangle has three sides both here, and on Alpha Centauri. We may choose to signal our intelligence to others by declaring our understanding of fundamental mathematical constants such as π: the ratio between the circumference and diameter of a circle. We ourselves have known of this ratio as far back as our written history penetrates; the ancient Babylonians and Egyptians were familiar with the concept, if not with the precise value of π. There is something appealing about the idea that we can broadcast abstract mathematical concepts, in the knowledge that whatever our differences in language or body form, whether we live on land or in water or in liquid methane, whether we are the size of humans, fleas or planets, whether we see with sight or sound or electric fields – there is no doubt that these mathematical principles apply to us all. This mathematics, therefore, would be instantly recognized by another species as a sign that intelligent life exists elsewhere in the universe.

But some philosophers have cast doubt on the idea that mathematics is the ultimate universal lingua franca.* For one thing, our understanding of mathematics is constrained by our very physicality. We are so used to the three-dimensional world that we rarely think of how alien the mathematics would be in a two-dimensional world. Ant-like creatures living on the surface of a very small sphere would find our mathematics very different to theirs. An ant could walk around its planet as if it were walking across a flat plane – although we could see that it was in fact moving on a three-dimensional ball. And in a world where you can only walk on the surface of a sphere

* See 'Communicating with the other', by Douglas A. Vakoch, in *The Impact of Discovering Life Beyond Earth*, edited by Steven J. Dick (ed.).

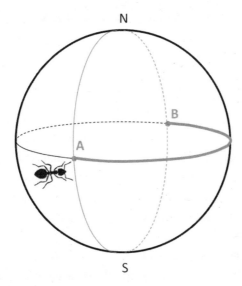

With different physical constraints, mathematics appears different. For the ant on the surface of this sphere, the circumference (going from A, all the way round the North and South Poles, back to A) is exactly twice the 'diameter' (going from A to B around the equator), and so π=2.

(no burrowing allowed!), π is not, in fact, equal to the familiar 3.14159265 . . . Consider a point on the equator of our imaginary ant planet, and the circle that goes through both North and South Poles. Our ant can walk along the 'circumference' of its world, going via North Pole and South Pole back to its starting place. But for the ant, the 'diameter' of the world is the path perpendicular to that polar route: along the equator to its most distant point. That line, going along the equator, is precisely half of the circumference of the planet, and so in this case π=2!

As humans, our particular intelligence evolved on the plains of the African savannah, to deal with the problems of the African savannah.* We can catch a tennis ball without solving

---

* See *The Blind Watchmaker* by Richard Dawkins.

Newton's equations of motion because throwing and catching come to us very naturally from generations of throwing spears and catching animals. But a blind mole living underground would find the concept of catching totally unfamiliar, and indeed might not comprehend that such a concept exists, until a mole mathematician capable of Einstein's abstract insight works out the equations of motion from first principles. Concepts outside of our physical experience are going to be hard for us to discover, and the physical experience of aliens is unlikely to be quite like ours.

As well as being constrained by our physical environment, the evolution of the science of mathematics on Earth has been driven by technological requirements: to build better temples (with walls at right angles to the floor), aqueducts (with arches to support their weight), catapults (and the ballistic trajectories of their boulders), as well as fighter planes and atomic bombs, with legions of scientists and engineers behind them. The trajectory of our mathematical discoveries has been shaped by our desire both to build structures and to knock them down with our propensity for war. A peaceful alien race may have no concept of ballistic technology, and one without religion may never have developed the technology to build imposing temples. Mathematical principles that seem to us fundamental and obvious may hold much less importance for aliens who have arrived at their state of 'intelligence' via a very different route.

But what of numeracy itself? Must all intelligent aliens count, for example? Even if they don't have fingers, or any such equivalent? How did mathematical ability even evolve on Earth, and is it likely to have followed a similar evolutionary pathway on other planets?

Evolution of mathematical ability

The ability to distinguish 'a lot' from 'a little' clearly brings tremendous evolutionary benefits to an animal. If you can head towards 'a lot' of food, and away from 'a lot' of predators, you will be more likely to survive. So it's no surprise that just about all animals of all levels of complexity have this ability. However, being able to tell the difference between a large pile of food and a small one is not mathematics. You might simply be able to detect that the smell is stronger from a large pile, for example, or that it occupies more of your visual field, leading to a stronger signal to the brain, saying 'go this way'.

The distinction between simple sensory response (understanding that the smell is stronger in one direction than another) and any kind of genuine numeracy (understanding that there are *more objects* in one direction than another) represents a major evolutionary step. Laboratory experiments have sought ways to test the true numeric ability of animals, and to disentangle numeracy from sensory sensitivity. As a result, several different species, from fish to pigeons to monkeys, have shown that many animals *can* perceive differences in *number*, rather than just differences in vague quantity. Although this is hardly something that we would call 'mathematics', it does form the essential foundation of mathematical competence. It is likely that no animal or alien can evolve a sophisticated mathematical ability unless it first acquires a basic numerical ability. Our own prodigious capabilities in algebra, calculus and statistics may only be possible because our ancestors could tell the difference between facing down two lions and facing down five.

The next step up from having a basic concept of quantity – something that we can see is evolutionarily advantageous – is to have an understanding of 'numbers' per se. This is a tremendous leap in intelligent ability and is much harder to explain evolution-arily. If I can distinguish between many lions and few lions, do I really need to know that there are *exactly* three lions? As such, the ability to recognize individual numbers is less widespread among animals, but far from absent. Early experiments showed that it was possible to train rats to pull a lever an exact number of times to get a reward: three times gets the reward, but four does not. The animals seemed to understand they must select a *specific* num-ber, rather than using a simple rule like 'more is better'.

Of course, we can be a little sceptical of the behaviour of rats in a maze: we have already seen that associative learning can be used to get animals to perform all kinds of feats, including skate-boarding. Is this likely to be some kind of by-product of intensive conditioning, rather like getting elephants to perform in a circus in the most un-elephant-like way? Although intensive condi-tioning can demonstrate that animals are, in theory, capable of performing 'intelligent' feats, it tells us little of whether or not they evolved those abilities to give them an adaptive advantage in the wild. Remember that our goal is to uncover the universal paths to intelligence, both among animals on this planet and also on alien worlds. It is the evolutionary purpose of intelli-gence that interests us, not the tricks that animals can perform to amuse us.

Other, more naturalistic, experiments have shown some very impressive numerical abilities in a handful of species, but only in a few. Chimpanzees can be taught to recognize written numbers and associate them with the quantities that they represent. This is a far more convincing demonstration of mathematical intelli-gence, but as they are our closest living relative it tells us relatively little about how widespread such capabilities may be on Earth, and by extension, elsewhere in the universe. Chimps are, after all, not *that* different from humans.

But some other, more surprising, animals *can* be numerate in the same way as humans, as in the case of Alex the African grey parrot in the 1980s and 1990s. Professor Irene Pepperberg at Harvard University took the biggest problem of researching animal intelligence – that we cannot ask the animals what they are thinking – and turned it on its head, quite simply by teaching Alex to talk.* Not just mimic human words, but to understand the concepts behind simple sentences, and to interact in clear, cognitive conversations at the level of a five-year-old human child. And the results were astounding. In a series of rigorous and carefully controlled experiments, not only could Alex correctly identify and explain the presence of objects of different colours, shapes and materials, he could count them. Not just a rote rendition of, 'one, two, three', but identifying the number of objects present, even where conflicting cues existed. For example, Alex was presented with a tray containing: one orange chalk, two orange pieces of wood, four purple pieces of wood, and five purple chalks (note that both colour and material overlap between the different objects). He was then asked, 'How many purple wood?' and answered correctly, 'Four'. This ability to interrogate an animal's cognition directly by asking it questions is remarkable, and exceptionally rare. Nonetheless, here was an animal with undoubted numerical ability. But what does this say about the evolution of mathematics on Earth?

Like the chimpanzee that could interpret written numerals, these species are exceptional in their cognitive abilities. They are, we have little hesitation stating, intelligent. But they are hardly representative of the myriad species of animals on Earth. Nonetheless, intelligent species – chimpanzees, parrots, us – have come to exist on this planet. It is at least *possible* for other animals to be this intelligent. But how did this arise, and what

---

* See *The Alex Studies: Cognitive and Communicative Abilities of Grey Parrots* by Irene Pepperberg.

were the evolutionary steps that led animals with goldfish-like numerical abilities (i.e. not much) to become mathematicians like us humans? When and how did the ancestors of humans become different in *method* from New Caledonian crows, developing an intelligence that was not so much 'greater' than that of their ancestors, but different in kind? And why? What kind of evolutionary pressure could have driven this change? Why do so few species – and yet, more than just one species – possess mathematical abilities? We need to know the answer to these questions if we are to assess the probability that similarly intelligent creatures – or at least creatures intelligent in a similar way – have evolved on other planets.

One possibility is that there is a significant advantage in having such intelligence – in which case the evolutionary force of natural selection acts very powerfully. But this seems unlikely, as neither Alex the parrot nor his cousins in the wild seem to make use of those mathematical abilities on a daily basis. Another possibility is that this kind of intelligence evolved in leaps and bounds, rather than the more familiar slow form of evolution whereby fangs gradually become longer and longer, or feather colours evolve to become brighter and brighter. Such leaps and bounds generally occur when some factor in the environment changes suddenly and dramatically, producing a strong need for animals to evolve rapidly to meet the demands of the new environment.

A sudden selective pressure such as this – for example a change in the climate in Africa – is likely what led our primate ancestors to adapt to living on the ground rather than in trees, leading to a very rapid adoption of our signature habit: walking on two legs. Similarly, it may be that our prodigious mathematical skills rapidly evolved as we became highly social, or perhaps as we evolved a language. If so, it is not unreasonable to think that animals on other planets will also not evolve mathematical intelligence until that time when suddenly faced with the need to do so, at which point one (or more) species will rapidly evolve to take advantage of the opportunities that exist for professional (or, perhaps more accurately, 'day-to-day') mathematicians.

As is clear from much of our own science fiction, when we discover intelligent aliens, we hope to discover intelligent aliens with technology. It is extremely unlikely that we will ever actually visit another inhabited planet outside of our solar system in my or your lifetime; the interstellar distances are just too great for any technology that we can envisage developing in the near future. So we hope that we will be able to send the aliens a technological signal, and receive a technological signal in return. Technology implies mathematics – or does it? Can we conceive of a technology without mathematics?

Humanity achieved some rather impressive technological achievements long before the invention of modern mathematics, particularly calculus 300 years ago, but we would never have made it to the Moon, invented computers or built radio telescopes without a good understanding of calculus and electromagnetic field theory. But perhaps a different intelligence could. For example, perhaps an alien equivalent of our own Earthling electric fish has such an intuitive understanding of what electromagnetic fields are, through its own everyday sensory life, that the workings of a radio would – to them – be obvious, without the need for a mathematical explanation! Just as we catch a ball without knowing the laws of physics, an alien electric fish may be able to build a radio without first discovering the laws of electricity.

This kind of speculative scepticism is useful, but somewhat unrealistic. Even if such a life form were possible, sooner or later it would encounter some natural phenomena which cannot be adequately understood without mathematics – remember that Earth's electric fish evolved their impressive abilities because of the absence of light in muddy river beds, and so they probably have a limited understanding of how visible light works. The underground mole is rubbish at catching a ball. Each species evolves capabilities to deal with its own environment, not to deal with all possible environments simultaneously. You can be smart, but it would be wasteful to evolve to know everything.

But if mathematics is necessary for technology, is it sufficient? That is, could there be other human characteristics that have shaped our technological intelligence, without which our technological achievements would not have been possible? What about curiosity? Philosophy? Literature? In our own history, our greatest mathematicians were also philosophers, and the development of science and philosophy often went hand in hand, albeit that this might well have simply been a coincidence of our own peculiar history. Can we conceive of a species that is capable of building a radio telescope, but that has no concept of poetry?

Interestingly, observing the spectrum of intelligent behaviour in animals, we can spot features of human behaviour that are not directly linked to the kind of problem-solving intelligence that is immediately understandable as being advantageous to the animals' survival. Sea lions, as well as intelligent parrots like cockatoos, enjoy music and dance – sometimes with excessive enthusiasm.* Chimpanzees laugh and appear to have an understanding of humour. These characteristics seem to arise from the animals' specific form of intelligence: sea lions and cockatoos enjoy dancing because of a fundamental biological connection between their social communication and their body movement. Chimpanzees laugh – quite likely – because of their complex social hierarchies, and the importance of expressing their pleasure and displeasure at the behaviours of others. If animals on Earth dance and laugh in response to their social interactions, there is no reason to doubt that aliens would do the

---

* For more information on the surprising behaviours of sea lions, visit the website of Dr Colleen Reichmuth: https://pinnipedlab.ucsc.edu/.

same. Behaviours that we think of as human innovations (like poetry or dance, for instance) are really just useful adaptive behaviours that have been co-opted for other, social, purposes. Therefore, there is no reason to doubt that in an alien society where body movements convey meaning, aliens will entertain each other by dancing.

As we have seen, intelligence appears to be a combination of two different abilities: a set of specific skills that evolved to predict specific features of the world around us, and a general ability that, whatever its origin, was leveraged to integrate together those specific skills into a far more powerful ability. If the evolution of intelligence has followed similar paths on other planets, we can say that intelligence is probably associated with a whole suite of abilities and characteristics, both prerequisites for, and inevitable consequences of, intelligent behaviour. And in that case, there is no reason to doubt that aliens will indeed have not just one specific ability that impresses us, but also have integrated together a suite of other capabilities as essential corollaries of what we call intelligence. Curiosity will drive philosophy, social interaction will drive art, and complex communication will drive literature. Really, these traits arise almost inevitably from the combination of intelligent skills that we, and presumably other alien species, possess.

Alien superintelligence

We usually assume that aliens will be more intelligent than us. Of course, any alien planet will have a diversity of life on it, some smarter, some less so. They will have communicative technological species like us, through a whole host of animal life forms of different levels of cognitive ability, all the way to the alien equivalent of jellyfish. But we often, and reasonably, assume that those aliens with whom we can expect to have a conversation will be more technologically advanced than us. It has only been just over 100 years since the first radio transmission made by our species; we are absolutely in the infancy of our technological development, and that makes it exceptionally likely that any aliens we encounter will be more advanced than us. They could be older than us or younger than us, but if we were to encounter an alien civilization at some random point in their history, there is only a tiny chance that we would happen to stumble across them in those first 100 years since they discovered radio. In the face of civilizations that could last millions of years, the chances of us being the smart kids on the block are minuscule.

All the same, length of tenure is no guarantee of degree of intelligence. They may be more advanced technologically, but does that mean they would be more intelligent? Imagine that the human race survives for another million years: clearly our technologies would advance, but would our mental capacities also? Is it the case that a species will continue to evolve higher and higher intelligence the longer that it exists, or might it reach a certain level of intelligence and go no higher? Popular science fiction certainly believes that any aliens we encounter will be super intelligent. But sci-fi depicts at least two distinct types of superintelligence: that which is essentially the result of technological advancements, and that which evolved biologically with the species itself. In science fiction terminology, this is the difference between a civilization that 'merely' has super-fast and powerful spaceships and a civilization that has 'evolved beyond' the need for such technology, and has perhaps evolved superpowers such as telepathy and telekinesis.

In the former case, one could imagine that after reaching a particularly advanced level of technology, an alien civilization (or even our own civilization) might transfer all requirements for intelligence to computers, which would leave organic biological organisms to occupy their minds with other pursuits. Perhaps we would be free to ponder the mysteries of the universe, free to philosophize, discover scientific truth and develop other intellectual hobbies. Or perhaps we would do nothing but play Tetris and watch internet cat videos; a choice between alien superintelligence and alien super-laziness. In the former case, we would not only have increased leisure time (and increased research time for scientists) because our technologies would have removed our daily struggle for existence, but our technology would also drive our scientific understanding, with bigger and better radio telescopes, faster computers and all manner of impressive scanners and detectors straight out of *Star Trek*. It seems likely that if we were to encounter ourselves 1,000 years from now, we would consider these future humans to be an 'advanced' civilization.

However, our biological intelligence would have remained broadly the same. We would be smart, yes, but essentially the same species. Robert Sawyer's excellent science fiction novel *Calculating God* tells the story of a technologically advanced and biologically very different alien species that visits Earth, and essentially engages in a long series of philosophical discussions with the human protagonist. For these aliens, technological advancement has not yet solved the mysteries of the universe.

But what about the second case, the possibility that an alien species might exist with intellectual capacities far greater than ours, by virtue of its natural biological evolution? Can we propose any realistic biological scenario under which this could happen? Indeed, is there even any need for natural selection to lead to super-intelligent adaptations much beyond those abilities that we ourselves already possess?

Animals on Earth have followed what is probably a very typical course: they need to predict the world around them. As such, they developed physiological and anatomical adaptations that allowed them to make predictions about the world using sensory information and some kind of processing apparatus: what we would call a brain. Any alien species that exploits a more unpredictable environment would have more challenging needs, and will develop more sophisticated, more adept, flexible and more accurate 'brains'. If intelligent animals have a social habit – which I think is quite likely, as we will see in the next chapter – then they are destined to evolve a language of some sort so that they can communicate the thoughts of their own brains to other members of their group. Taken to its logical conclusion, it is likely that technology will eventually develop.

Once a species is sufficiently technologically adept, it will learn how to construct brains more powerful than its own, some equivalent of artificial intelligence (as discussed in Chapter Ten). This is close to the situation in which we find ourselves now, or at least will do in the next 100 or 200 years. From this point, we may well develop intellectually as individuals and as a society, but the evolutionary pressures for our biological intelligence to increase as a species would have disappeared. Why evolve to become more intelligent if computers can do all the work? The evolutionary pressure to become super intelligent will have evaporated.

But what if an intelligent species evolves that is not social? I think it is doubtful that technology is possible without sociality; one individual, no matter how intelligent, simply cannot build a spaceship or a computer on their own (who would pass them the spanner?). In this case, as long as their environment still provides them with challenges that can be solved better by greater intelligence, such organisms may continue to evolve bigger, more complex, more reliable brains. Such a route to superintelligence seems at least a possibility, albeit a highly unlikely one. Fred Hoyle's novel *The Black Cloud* portrays this sort of solitary intelligence drifting through the universe on its own, but with abilities far greater than any human-like species could have evolved even given an incredibly long period of evolution.

Hoyle's Black Cloud is profoundly biologically unrealistic. A prerequisite for continuing selective pressure on intelligence is that the organism continues to be challenged by its environment with problems that can be solved by greater and greater intelligence. It is hard to conceive of an ecosystem where unlimited intelligence continues to provide practical solutions to the problems of day-to-day life. Sooner or later, you run out of existential problems to solve. In fact, as with many science fiction super-intelligent aliens, the Black Cloud's intelligent ability appears to be an end in itself, rather than a means of improving its evolutionary fitness. As we have discussed, evolution does not seek out ends in themselves, only relative improvements on an organism's current abilities. Unfortunately then, the concept of super-intelligent aliens floating through the universe philosophizing for mere intellectual pleasure, although appealing, is not biologically plausible. So true biological superintelligence evolving from the continuing challenges of the environment seems unlikely. Either the species will develop technology instead of better brains, or the species will run out of intellectual challenges.

However, there is another mechanism to the evolution of true superintelligence. In this scenario, multiple individuals are bound together mentally in such a way and so tightly that their thought processes are shared almost instantaneously and completely. Like a supercomputer composed of multiple smaller computers working in parallel, such a colony of intelligent beings could indeed be perceived as being a single super-intelligent organism. There are of course many parallels for this in the natural world. Many creatures live in colonies, hives or even temporary aggregations which appear to have an intelligence of their own, far beyond the abilities of their individual members. One of the most visually impressive examples is that of schooling fish. Each

A school of fish evading the attempts of predators to attack it.

individual fish swims in a direction that has been shown to be governed by quite simple rules based on the direction and distance of its immediate neighbours. Put hundreds of such fish together, however, and the school as a whole appears to behave intelligently. Predators attempt to rush into the centre of the school, which divides, almost magically, leaving the sharks or dolphins empty-mouthed. The fact that an aggregation of fish can show such adaptive and seemingly intelligent behaviour, when each element is incapable of doing so on its own, is a simple example of emergent superintelligence: the whole is more than just the sum of its parts.

Another example of emergent intelligence is found in the hives of honeybees. When a colony needs to move to a new hive site, scouts fly out and investigate possible real estate options. Each one returns to the current hive and communicates to those around her the benefits of the site she has discovered. The hive as a whole is faced with two problems: multiple scouts may be 'recommending' different destinations, and each scout can only communicate to a small number of bees, not the entire hive. As it would be disastrous for the colony to fly off in different directions, some consensus must be reached. But how? There is no one chief bee that makes a decision. Again, simple rules dictate complex behaviour. If a scout provides a very promising recommendation of a particular site, she will convince many other bees to follow her for an additional viewing of the property. Each of those bees will return and make their own recommendations, and in this way information about all the available sites is integrated into what can be considered (in every meaningful way) the 'brain' of the hive. Except that, rather than being a bodily organ, this brain is a collection of individuals, each only communicating to a few others (much as neurons in our own brains are only connected to a few other neurons). Competing destinations are competing for the attention of this hive brain, and eventually a tipping point is reached, the entire hive converges on a single consensus, and off they fly.

Although we think of colonies as being composed of separate individuals, each with their own interests, and own processing abilities, it is important to remember that our own bodies, and those of every other animal on the planet, are in fact the result of a series of opportunistic cooperative associations. When multi-cellular organisms first evolved on Earth, they too needed to communicate with other single cells in the growing colony. Today, the cells in our bodies are in communication with each other so completely that we consider ourselves to be a whole organism, rather than a collection of independent parts. By furthering this analogy, it is certainly conceivable that a single super-intelligent organism could evolve by the tight association of many intelligent creatures, bound together to such an extent that they can no longer properly be considered individuals.

Although a popular trope in science fiction, an alien organism composed of such super-cooperating almost-individuals is a highly constrained possibility. Equivalent creatures on Earth, such as the Portuguese man o' war we saw in Chapter Four, are tightly bound colonies of individual animals called zooids, no matter how much they may look like a single organism. But the man o' war is very simple in both its behaviour and its structure. The primary constraint on the complexity of hive organisms is the question of how much information can be transferred between the individuals, and in the case of the zooids of the man o' war, the answer is: not much. True hives, such as bees and ants, are much more complex, and their communication is correspondingly more complex. But ants and bees are also genetic clones, meaning that the individuals aren't really – in evolutionary terms – separate individuals. A true hive intelligence, similar to the fictional Borg aliens from *Star Trek*, would require a highly complex and information-rich communication channel linking the individuals – and that is indeed exactly what the science fiction authors postulate. But could such a system evolve on its own? It would seem that this is far more likely to occur as the result of conscious engineering – and this we will discuss in Chapter Ten.

I would not claim to have arrived at a universal definition of intelligence, and it may be that no such definition exists. But thinking about the different kinds of intelligent life on Earth, we can certainly identify some specific predictors that should be common to intelligent life across the universe. All animals sense their environment and react accordingly to solve problems. What we will call 'intelligent animals' use multiple sensory sources and integrate that information using a crucial process called learning. Learning isn't intelligence in itself, but it is the mechanism by which all the spectacular and specific cognitive abilities of animals are brought together to make something bigger.

If learning is so useful, it will be found on other planets, as will specific intelligent skills. On a planet with exceptionally hard-shelled but tasty nuts, animals will evolve intelligent skills to open those nuts. Specific intelligence is just another trait, like long fangs or camouflage skin, that will evolve if it provides a fitness advantage.

And so, given that both general and specific intelligence are incredibly likely to evolve everywhere, under what conditions will creatures integrate the two, combining their abilities into something that resembles the intelligence with which we are familiar? Pretty much under *any* conditions. Given the number of animals that can learn and integrate between specific intelligent skills – dogs, crows, dolphins, octopuses and myriad more – it is inconceivable that we are talking about an ability peculiar to planet Earth. Evolution will favour intelligence as we know it.

Sociality and technology (which can be as simple as manipulating a twig to fish out a grub) seem to be both evolutionary requirements and consequences of intelligence; the relationship between these abilities is so hand in hand that it makes little sense to say which came first. But the interplay between them probably has a crucial role in the evolutionary mechanism for intelligence. And the logical end point of intelligence is either with external brains, such as computers, or with evolution continuing to push the individual's biology until it becomes what we would call 'super intelligent'.

From considering the intelligence of individuals, it would seem that social interactions are crucial in driving the evolution of complex behaviour, communication and abilities. Sociality, therefore, is the subject of the next chapter.

7.   Sociality – Cooperation, Competition and Teatime

Are aliens social? This is perhaps the most important question that we ask in this book, and not only because we'd rather like to sit down with them and have a cup of tea. It is important because whether or not aliens have a social nature determines whether they have advanced technology (including radio telescopes and spaceships), how much we would have in common with them and whether we would be able to relate to their thoughts, desires and fears. If they have alien families and alien jobs, as well as alien pets and alien supermarkets, how different can they be?

To explore this we'll ask the question, 'In how many different ways is it possible to become a social species?' Is it a foregone conclusion that social animals will exist on any inhabited planet, or is it going to be a fluke of circumstance? Is there something peculiar about Earth that leaves us surrounded by social species, from ants to zebras? Or is sociality something that can be deduced from first principles, from theory? Beyond that, do the diverse range of social animals on Earth illustrate *all* the different ways that sociality could evolve, or might there be utterly unfamiliar kinds of social organization on other planets? The former seems likely: evolutionary theory tells us how and why sociality can evolve, and what theory predicts is what we see on Earth. In which case we are in a good position to make predictions about the fundamental structure of alien communities, and therefore about the nature of the aliens themselves.

There are three important questions we need to address in this chapter. Firstly, why do animals live in groups, and why do they actively cooperate in those groups? Secondly, what are the conditions under which animals will evolve to form cooperative societies, and the constraints that might prevent animals from cooperating? And thirdly, what are the outcomes and the expected evolutionary consequences of animals living together in this way? When hoping to draw conclusions about what aliens might be like in their social characteristics, we must again turn to the single uniting mechanism of biological evolution: natural selection. We must be careful not to draw too many conclusions from the idiosyncrasies of human sociality, nor from sociological research on our own species. Instead, we hope to take many steps backwards to the most basic, most fundamental evolutionary processes that must drive animal behaviour everywhere in the universe. Understanding sociality from an evolutionary perspective will make our predictions about alien behaviour much more accurate than just extrapolating from human experience.

Why live in groups?

So we begin with our first question: *why* do animals live in groups? Our first insight into this goes back a long way, long before and far removed from the animal world: even bacteria live in groups – and they're not even animals! But more than just *living* in groups – which may not be surprising, because most bacteria aren't very good at moving away from wherever they start out – bacteria actually *cooperate* in groups. Microorganisms often secrete slimy substances that form a thin, sticky layer that protects the reproducing organisms from the environment. As the bacteria multiply, this slimy coating is so thin, and space inside is so limited, that the organisms within are always interacting. They compete for nutrients and, occasionally, kill each other by secreting toxins to make more room for themselves. It's a veritable microscopic Gotham City. Despite this, some bacteria cooperate by secreting chemicals that help other cells, rather than harm them; for instance digestive enzymes that make food available to be absorbed by anyone in the vicinity. And the members of the colony also communicate; exactly which particular helpful chemicals a bacterium produces will be affected by the concentration of special signalling chemicals generated by other members of the colony. Living in a bacterial group obviously serves the organisms well.

On the face of it, the advantages of living in groups are obvious. Social animals are better protected from predators; they can find food in ways that would be impossible alone; and cooperation allows animals to achieve spectacular feats of construction, from rabbit warrens to termite nests and skyscrapers. Mates are nearby when you live in a group, and although it's possible to say that *all* animals on Earth are social to some extent, because they have to come together to mate, we can't be sure how aliens 'do' mating, so this may or may not be a universal advantage of group living.

There are also some less obvious advantages to sociality, which we'll explore in more detail later in this chapter. You can provide support to members of your family, and perhaps it is simply the default option when leaving the place you were born is just too dangerous to do alone. If you cannot move at all, or only with difficulty, as in the case of microbes and plants, sociality may be something you simply have to learn to live with. Even stereotyped millennials are forced into a kind of sociality by not being able to afford moving out of their parents' basement.

But if sociality has so many advantages, why do only some animals on Earth show feats of spectacular sociality: endless wildebeest herds, complex termite nests, New York, and so on? Others – tigers, polar bears, sloths – are essentially solitary. Because, of course, being social also has its disadvantages. With many individuals in a group, there is increased competition for resources, there is scope for conflict and injury, and there may be additional disadvantages, like being more prone to parasite or disease transmission. So clearly there's a trade-off, but what determines the nature of that trade-off? Under what conditions does sociality tend to evolve and, critically, how universal can our conclusions be about the nature of sociality, and the constraints that drive it to occur?

Avoiding predation is one advantage that springs to mind when it comes to social animals. Zebra live in herds and musk oxen form defensive circles to protect themselves from predators. Baboons actively harass and chase off leopards, while rabbits dig communal tunnel systems as protection. Many animals, like meerkats and scolding jays, have alarm signals to warn others of predators. Predation is universal, because no ecosystem can exist for long without someone trying to take a bite out of somebody else; the selective pressure on acquiring as much energy as possible is just too strong. Even top predators like lions and orca are vulnerable to becoming someone else's food at some stage in their lives. So we can be confident that alien worlds will (much to the delight of Hollywood) be full of voracious predators.

Furthermore, predation isn't just inevitable, it is one of the most powerful drivers of adaptation. Faced with the possibility of being destroyed and therefore unable to beget offspring, the evolutionary advantage of avoiding being eaten is huge. Every animal – and alien – must evolve ways to avoid becoming the victim. On the other hand, predators will die without eating, and so they also have huge evolutionary pressure to become better predators. Better predators drive better anti-predation defences; and better anti-predation defences drive better predators. This positive feedback – as we discussed briefly in Chapter Four – is known as an evolutionary arms race, and arms races are responsible for some of the most impressive feats of animal behaviour. For example, there are ants that commit suicide by exploding themselves to protect their nest from invaders, and moths that produce acoustic countermeasures to jam the sonar of predatory bats. Just about anything goes in the battle to avoid being eaten.

However, just as the nuclear arms race between the United States and the Soviet Union ate up an increasingly unsustainable amount of resources, similarly in the animal world an arms race cannot continue forever. Anti-predation countermeasures make the life of a predator more difficult, and effective predators place a huge burden on their prey's life. If you spend all your time seeking prey, or looking out for predators, you don't even have time to eat or mate! Known as 'non-lethal predator effects', this is a more important constraint on the behaviour of most prey species than actually being eaten. If you've ever watched a deer or a rabbit nervously foraging at the side of a field, you'll know how imposing the fear of predation is. Running off as soon as you see a potential predator doesn't bode well for your calorie intake.

Something has to give. The iconic face-off between *Tyrannosaurus rex* and *Triceratops* illustrates this well (whether or not this scene ever actually took place in reality): super-vicious predator facing super-protected prey. Investment in increasingly ludicrous armaments becomes unsustainable, so this escalation can't just go on forever.

Tyrannosaurus vs *Triceratops*. The iconic painting by wildlife illustrator Charles R. Knight.

When faced with such efficient predators, rather than holding up a massive horned skull, perhaps living in a group may be a better way of keeping safe. In the first instance, simple statistics indicate that being part of a group means that you as an individual are less likely to be the unlucky one picked off by the predator. Herding, or diluting yourself in a group of others to avoid being eaten, seems an inevitable consequence of the situation where animals in large numbers and open spaces are chased by predators. Who wouldn't dive into a group of other animals to hide from a rampaging *T. rex*? Herding should exist elsewhere in the universe because it is a straightforward passive process, driven by simple calculations of how to reduce your own risk. Of course, not all animals herd – if you are smaller, more agile and faster, you may stand a better chance on your own. If your predators ambush you in dense forests, herding is less of an advantage. But the principle of herd dilution can in itself be enough to set the stage for animals to begin to form social groups.

There are other ways, too, that predation could cause animals to aggregate, but unlike passive herding, these adaptations require that the animals actually *cooperate*. For example, sentinel behaviour and alarm calling is very widespread. Having someone else on the lookout means that you don't need to spend as much time looking up yourself – you can concentrate on eating. Meerkats sit up on their haunches from time to time and scan for predators while everyone else is foraging, alerting the whole group if any danger is detected. One of the animals I've studied extensively is the rock hyrax, a small furry creature that looks like a very large guinea pig, but is actually more closely related to the elephant. The three species of hyraxes alive today are the last surviving members of a group that once included animals as large as horses. Hyraxes come out from their dens to forage as a group, but some individuals always take position on a prominent rock and, rather than eating, watch for predators and give a

Hyraxes (left) and meerkats (right) on sentinel duty.

loud alarm call if any are seen. Everyone then rushes back to cover.

Sentinel behaviour is very different from passive herding, and the animal on guard has paid a very real cost for the sake of his or her colleagues. Not only does the sentinel miss out on feeding opportunities, but they are placing themself in the most prominent, most obvious position to attract the attention of predators. Why should a sentinel perform such an altruistic act? Natural selection works to favour your own survival, not that of others. Nonetheless, the behaviour has evolved, and therefore it must be favourable. What drives the evolution of these seemingly altruistic strategies?

With predators, we also see both simple aggregation and also altruistic cooperation. Wolves, for example, cooperate (among other things) to bring down prey larger than themselves because none of the individual animals could survive otherwise. By hunting together they receive only a share of the food, rather than the entire animal, but that's better than nothing, which is what they would get if they hunted alone. But predators also show what appears to be truly altruistic behaviour. Lionesses care for the cubs of other females. Meerkats catch prey to feed the young that are not their own babies. Wolves themselves care for and bring food to regurgitate to pups that are not their own. These are behaviours that cannot be explained simply by passive aggregation; they reflect a fundamental mechanism by which a group of animals – Earthling wolves or alien wolves – do better as a group than as individuals.

Many, many animals – both predators and prey – live in groups and help one another. Seemingly altruistic behaviour is widespread on Earth. For some species, altruistic social behaviour is almost their most defining feature – from bees in a hive to humans in the welfare state. But altruism is, on the face of it, such an odd way to behave that it calls into question the generality of our assumptions about animal behaviour. Is the altruistic social behaviour that we see in wolves, humans and hyraxes the result of some weird phenomenon peculiar to Earth, and not to be found on any other planet? If so, then perhaps life is common in the universe, but sociality is very rare. And if sociality is rare, perhaps we cannot expect aliens to have accomplished technological feats (e.g. building a spaceship) that require a high degree of cooperation. Clearly, arriving at a very fundamental understanding of *altruism* is essential if we want to know how our equivalents on other planets behave.

The existence of altruism puzzled early evolutionary biologists. The sentinel eats less, and may be more likely to be eaten themself. Why not refuse to do sentry duty, and let someone else take the risk? Digging a burrow requires energy – why not let the others do the work, and then take the fruits of the communal effort? Cheating and freeloading seem to be a very effective way to get what you want, and in the cold actuarial accounting of natural selection, selfish strategies should always win out, so that individuals that cheat will leave more offspring and their cheating genes will spread throughout the population.* Crime pays.

---

* *The Selfish Gene* by Richard Dawkins is essential reading, and the gold standard in accessible explanations for evolutionary behaviour.

One solution proposed by some early biologists like Konrad Lorenz in his 1963 book, *On Aggression*, was that there might be some kind of vital force that causes animals to act 'in the interest of the group'. This was a controversial idea for a while, but as there was no real mechanism by which this group interest could work, such a vague semi-explanation did not hold force in the scientific community for very long. Rather, there *are* rigorous answers to this conundrum, and the generally accepted explanations for altruistic behaviour in animals rely on sufficiently broad and general mathematical principles that we can certainly extend their presence – or absence – to other planets. The two most important factors driving the evolution of altruism are particularly relevant to us, because they are likely to be acting on any alien planet as well.

How do groups stay together?

Understanding how altruism has evolved on Earth is the first part of the answer to our second question: what makes animals cooperate. The key insight is that animals living in groups tend to be related to each other, and so share many of their genes. In a group of meerkats, many of the female members are sisters of the dominant female, and so the babies are their nieces and nephews. These subordinate females don't have their own offspring, because the dominant female doesn't allow it: generally by bullying tactics, evicting pregnant sisters from the group, or simply by killing the babies of subordinates. But for the subordinates, by helping the babies of the dominant, the aunties are helping some of their own genes to succeed. Admittedly, not as many genes as they would share with their own offspring – if they had any, which they don't, because they don't breed – but these young meerkats are hardly strangers either, they are kin. The subordinate females have a choice: help their nieces and nephews to survive, or give up having any genetic representation in the next generation at all.

The mathematical basis of what is now called kin selection (as discussed briefly in Chapter Two) is fairly straightforward, even if many new details are still being argued over. Essentially, kin selection means that the effort I make to support my relatives should be proportional to how related to me they are.★ In a possibly apocryphal story set in a London pub, J. B. S. Haldane summed it up in his typical pithy but tongue in cheek style: 'I would gladly lay down my life for my two brothers . . . or eight of my cousins.' Kin selection has revolutionized the study of animal behaviour because, even if it cannot directly explain *all* cooperation, we can't start to explain *any* cooperation without it. Wolf packs, lion prides and hyrax colonies all have an over-representation of related individuals, and this must influence the decisions that the animals take. Favouring your genes in other individuals is evolutionarily advantageous and, conversely, any tendency to ignore relatedness when deciding whether or not to help others would carry an evolutionary cost.

Kin selection is most strikingly observed in social insects like bees and ants. In these species, the advantages of living together are incredibly powerful. Not just to give communal protection, but also to make foraging more efficient, for instance the bee's waggle dance that indicates the location of food. Some ants too are abso-lutely dependent on their communal lifestyle for food. Eating the tough and indigestible cellulose of plant material is not very nour-ishing, so some ant species collect leaves and grasses, and bring

---

★ This principle is known as Hamilton's rule. I share 50 per cent of my genes with my children, the rest come from my wife. If I give them £100, and they can make more than £200 return, that's more than £100 for my genes that sit in my children, and so a more worthwhile investment for me than keeping the £100 to myself (with whom I share 100 per cent of my genes).

them to vast underground chambers, where other individuals farm fungi that decompose the cellulose and make the food more digestible. Agriculture is not just a human invention. Maintaining and protecting these fungus farms is a communal effort, so the species simply couldn't survive as isolated individuals. Some insects also have a perplexing trick up their sleeves: their familial relationships are utterly different to ours. Females are 50 per cent related to their parents, and to their daughters, just like us. But males hatch from unfertilized eggs, and so have no father – they share 100 per cent of their genes with their mothers. The surprising result is that sisters – who generally make up the bulk of the colony – are related to each other not just by 50 per cent like in humans, but they actually share 75 per cent of their genes with each other. These females are 'super-related', and so the pressure of kin selection to cooperate with each other is correspondingly stronger.

Now, it is not yet clear whether this genetic effect is indeed the reason why these species live in such highly communal societies, but it is clear that the precise nature of relatedness drives the behavioural decisions that animals make and influences the kind of sociality that they follow. In fact, a species composed of nothing but clones – rather like those bacterial biofilms with which our story began – would be highly cooperative. There is really no evolutionary reason for one individual to preserve its own life over that of its multitudinous identical twins. An extreme example is multicellular life, like humans – your liver cells have nothing to gain by turning against your blood cells! Suicidal exploding ants are also acting out an instinctive game of promoting the genes they share with their colony mates, even if that means their own lives coming to an abrupt end. Alien species where relatedness is, for whatever reason, extremely high would almost inevitably be extremely social. But how can we know whether aliens will be very related to each other or not? Knowing about alien 'families' means knowing how aliens get 'in the family way'.

Birds and bees and aliens

Unfortunately, we have painted ourselves into a difficult corner with our narrative about inferring alien life from Earthling examples. Sociality, this vital aspect of our behaviour that we would hope to find in alien creatures too, depends heavily on relatedness, and therefore on sex. But about alien sex, we can say very little. We don't know if aliens even have sex, or what it would be like. Our precise biochemistry defines which molecules (like DNA) determine our genetics, which in turn defines our inheritance, and our sex lives.

But our biochemistry is something specific to Earth. DNA and its sister molecule, RNA, seem to have emerged right at the beginning of life on Earth – possibly as the result of a huge chemical coincidence. It is far from certain that DNA is the only molecule in the universe that can be used for 'genetics'. The structure of DNA determines how our chromosomes work to transmit traits from parents to children. Animals on Earth almost always have two parents, because animals get one set of chromosomes from a mother and one from a father. So what if alien life forms do not use DNA? Life arising on another world based on a different biochemistry might have different constraints on how parents beget children. What if aliens have three or four or five parents each? Or more than two sexes – or maybe no sex at all?

However weird alien sex is, we *can* say that, as long as (like on Earth) each generation is less and less related to its ancestors, kin selection should be a driving force for sociality on alien worlds – and then surely it will lead to altruism and cooperation just as it does here. True, the precise effects of kin selection depend on the precise nature of inheritance, and this we cannot know. Just the slightest shift of inheritance relationships from 50 per cent to 75 per cent that we see in social insects corresponds to hugely different social structures. We mammals can't even begin to imagine what it would be like to live in a hive of clone boys and super-related girls. Vastly different heredity – 'genetics', if we can call it that – in alien life would mean vastly different rules of inheritance, and therefore the outcome of kin selection is almost unpredictable. But as long as aliens are less related to their grandchildren than they are to their children, some form of sociality should evolve.

Fortunately, kin selection is not the only universal rule that drives social behaviour, and so we still have some tools in the bag that can help us to predict the presence, or otherwise, of alien societies.

Firstly, sociality may be a by-product of other constraints that the animals face. For example, if it's a dangerous world out there, it may just not be worthwhile for children ever to leave their parents' home. At the very least, they might want to stick around for a while before heading out into the wide world. European rabbits live in communal warrens (although they don't help raise each other's babies), and subordinate animals (both male and female) put up with exceptional bullying and violence at the hands of the dominant animals. The naturalist Ron Lockley gave a fantastic and intimate view of rabbit behaviour – including the gory violence – in his book *The Private Life of the Rabbit*, which was the inspiration for Richard Adams's *Watership Down*. However, despite the widespread violence, it's more dangerous outside of the warren, and independent survival is simply not an option for rabbits. And as they say, where there's life, there's hope – perhaps a subordinate male will be able to sneak in a mating here and there, or will succeed the dominant male when he dies. Sociality can arise if need be, even in animals that don't seem like they should be social.

Alternatively, as a female zebra, you may be forced to stay in a group by a dominant male, who prevents females from leaving his harem. These simple mechanisms leave animals with no choice but to be social, and so drive the association of animals into groups. But merely associating isn't the same thing as co-operating; coercion is unlikely to drive cooperation, and certainly not anything like the complex and technological society that humans have achieved.

To understand the other mechanisms of cooperation, we need to consider that, whereas altruism is a behaviour whereby one animal sacrifices something for the benefit of another animal, a win–win situation is also possible, where both animals benefit. True, it takes energy to build a burrow, but the investment is the same for every rabbit, and every rabbit gains the same protection from predators and from the weather. Surely such mutual cooper-ation should be commonplace? Let's work for the benefit of society, and then everyone gains!

Indeed, this mutualism *can* exist even without the benefit of kin relationships – but only under certain circumstances. If you remember the 1970s sitcom *The Brady Bunch*, you can see how that works. Two parents, each with their own offspring, cooper-ate to raise their (unrelated) children together. No assumptions about alien sex needed.

It was the work of the mathematical economist John Nash★ and others that gave us the theoretical framework of game the-ory, and through this, a fundamental understanding of how it is that animals sometimes do and sometimes do not cooperate. Game theory is a beautifully simple mathematical way to predict how someone (animal, human or alien) will behave *in the long run*, in a situation where other individuals are also trying to get the best deal for themselves. Like Isaac Asimov's fictional mathemat-ician Hari Seldon in the *Foundation* series, we may not be able to predict individual decisions, but as evolution works over long time scales, game theory predicts which behaviours species are

---

★ Subject of the Oscar-winning film *A Beautiful Mind*.

likely to evolve, and in particular the way that they will cooperate or compete.*

This problem is often illustrated using two example scenarios: the Hawk–Dove game and the Prisoner's Dilemma.† Despite these rather obscure names, each scenario is beautifully simple and widely applicable across species and fields of endeavour, from bacteria to humans, from mating strategies to world diplomacy.

The Hawk–Dove game works as follows. Consider a population consisting of perfectly cooperating and peaceful individuals (Doves). Let's say it's a population of deer, where the males mate with females as and when they wish, without interference from any third party. A kind of fuzzy Woodstock. If a single selfish individual (labelled a Hawk, even though in our example it's a deer) with an aggressive penchant for monopolizing the mating is introduced into this population, he will ruthlessly exploit the others and be far more successful (as they say in country music: all taking and no giving). His selfish genes will spread through the population, because animals with those genes will prevent the other males from mating. The peaceful hippy males don't get a look in, and eventually natural selection ensures that everyone is a selfish Hawk.

However, a population of aggressive male deer is a violent place to be. If everyone is trying to monopolize mating, there is conflict and possibly injury. Males grow antlers to discourage others from fighting with them, and to use them as weapons if necessary. This

---

* The *Foundation* trilogy proposes that human history, the rise and fall of civilizations, even galactic empires, can all be predicted given sufficiently complex mathematics. Of course, the details of history cannot be predicted – there are too many random events influencing human behaviour – but in the long term, subtle but clearly defined patterns emerge.

† For a much more detailed description of these and other game theory concepts, see John Maynard Smith's *Evolution and the Theory of Games*, or Dawkins' *The Selfish Gene* for a more general introduction.

is, it would seem, a 'worse' place to live than the peaceful commune with which the deer started. But consider that in a population of fighting male deer, a single peaceful (Dove) individual has a certain advantage. While all the other males are busy wasting energy growing huge horns and fighting with each other, the peaceful male sits back and saves both energy and injury relative to the others. He may be able to sneak in and mate here and there while the other males are busy challenging each other. Perhaps Doves can eke out a living in a population of Hawks after all.

The key insight here is that the success of an individual – its evolutionary fitness – depends not only on its own traits, but *also on the behaviour of the rest of the population*. If everyone is selfish, it's good to be peaceful. If everyone is peaceful, it pays to be selfish. No one strategy is unequivocally the best; it depends on the strategies of others, and there's an advantage to being different. Nash's insight was that there might be an equilibrium strategy – part selfish and part peaceful – that cannot be bested by any other strategy. When the population is, for the sake of argument, 20 per cent Dove and 80 per cent Hawk, anyone who adopts a mixed strategy of being Hawkish more than 80 per cent of the time *or alternatively* Dovish more than 20 per cent of the time is *worse off* than everyone else. The mix that includes both behaviours at a particular ratio is stable, because at that ratio any individual acting with a more extreme strategy is losing the advantage of being different.

This Nash equilibrium, or as it is often called in biology, evolutionarily stable strategy, explains how animals might cooperate because it provides them with greater fitness, even though it would seem as though they would do better by simply being selfish. Nash equilibria are a fundamental outcome of game theory; they are in no way exclusive to reproductive systems, mating, relatedness or sex, and so we can confidently expect the rules surrounding these behaviours to be the same here and on every alien planet.

The second theoretical game often talked about is the Prisoner's Dilemma, and this also illustrates how the evolutionary outcome of cooperation and competition may not always be what you first expect. In the Prisoner's Dilemma, two criminals are being interrogated separately and incommunicado by the police, and are facing a long jail sentence. Offered a more lenient sentence if they squeal on their co-conspirator, what should they do? If one keeps their mouth shut but their colleague grasses on them, the 'honest thief' is in big trouble. But if they both keep silent, they'll both get off scot-free – clearly the best option. Of course, neither knows what decision the other will make. A simple mathematical analysis, however, shows that in the long run, the best strategy is to squeal. In other words, although we may know that the best thing for everyone is to cooperate (don't say anything and the police don't have anything on us), it is nonetheless a riskier option (if your partner squeals you then face the worst-case scenario), therefore prisoners will still choose to betray their colleague (you end up in prison but are guaranteed a shorter sentence). The *best* solution (cooperation) isn't necessarily a *stable* solution.

A similar effect is observed in nature, most especially in raising young. Clearly, a parent wants their offspring to survive, and survival is best guaranteed by both parents working together. But if I stay in the nest and look after the chicks, perhaps my partner will desert and leave me to raise the kids on my own. Better to take the initiative first and fly off to mate with someone else, rather than risk being the chump myself. This behaviour is surprisingly common in birds, from kites to tits, and possibly occurs in as many as 40 per cent of bird species. On the face of it, deserting your mate may seem to be contrary to natural selection, i.e. choosing the less profitable course of action – because of course it's best for both parents to work together. But evolution favours long-term outcomes, and game theory predictions are similar to human bet-hedging: I am prepared to accept a lower return on investment, in exchange for a more reliable investment. In the real world there are conflicting factors influencing whether there is an advantage to deserting, or an advantage to cooperating. Perhaps I will have to deal with the same individual again in the future (I don't want them to be mad at me), or perhaps if everyone is playing safe it is worthwhile for me to make a risky investment. But these are just additional complications to the equations – and the equations themselves are universally applicable. Natural selection is driven by fitness, and fitness depends on the games you play with other individuals – both here and on other planets.

Where does sociality lead to?

As we have seen, we can say much about whether or not animals will cooperate on other worlds, even if the way that aliens are 'related' to each other is very different to our family relationships. We can expect more similarity if relatedness on their planet is anything like ours. But let's return to our third question: what will be the outcome of this cooperation? Can we draw conclusions about the behaviour of alien animals by looking at behaviour here, or are there surprises waiting for us in outer space? Might cooperation lead to unfamiliar behaviours and societal structures, quite different from what we see on Earth?

We have already talked about the importance of cooperation for protection against predation, in terms of active defence like alarm calls and physical resistance, as well as passive defences like burrows and other nest structures. Cooperation also helps animals to find food for themselves – through sharing information about the location of food sources, for example – and helps predators become more efficient by hunting as a group. But the effect of hunting as a pack isn't just that animals are more effective at finding food; pack hunting also serves as a basis for a social group that cooperates in many other ways as well. Wolves don't just cooperate to hunt, they cooperate in raising young and in defending their territory from other packs and predators such as bears. Once cooperation has been established, for whatever reason, consequences follow. Cooperation begets cooperation. In particular, the requirement to live in a group (because they need to hunt together) means that they must also *get along* together. Social skills must be developed.

Sociality has the potential for being hugely beneficial – protecting individuals, gathering resources more efficiently, developing technologies – but it comes with costs: the sharing of food, for example. If wolves were to kill an elk, and then to kill each other fighting over the carcass, not much benefit would accrue to any of them. Therefore, social animals have a range of social skills, a set of signals and interactions that allow them to keep the peace and determine each other's intent before making any costly faux pas. Those of us with pet dogs can become acutely aware of these complex interactions. As we don't share a language with our dogs, we don't have the ability to ask an animal outright, 'Can I stroke you, please?' We must read the messages that they are sending: expectant perked ears, wide, scared eyes, a play bow, or tail between the legs. These are signals that evolved, not for our domestic benefit, but because the only way that animals – competing and cooperating at the same time – can live together is if they can anticipate the responses of the others. It is a strong argument (although I am not aware that this has been tested either in theory or in experiment – yet) that *any* social species *must* develop social signals, because of the inherent conflict that exists in all cooperation: when you give of yourself to help others, you lay yourself open to exploitation.

Another trait that seems to arise almost spontaneously in cooperative animal societies is a sense of reciprocity: I'll help you if you help me. This is another kind of behaviour that requires some considerable explanation; why should I not just accept your help and then cheat on you, not returning the favour? That seems like a better deal for me. The obvious answer – to us as humans – is that if I'm known to be a cheater, no one will help me next time. But most animals don't have the cognitive ability to remember each individual in their group, or to count up the times that each one helped and the times that they cheated. Yet incredibly, reciprocity exists in many animal groups, even when relatedness is low and where cheater recognition seems not to occur. Reciprocity seems a good way to cooperate, but how does it evolve?

The most famous example of reciprocity in animals is the fascinating and much-maligned vampire bats. These tiny flying mammals expend so much energy in flight that if they fail to find any food in one particular night, they are in real danger of starving to death. When they return to their roost, the animals that have succeeded in feeding offer a bit of regurgitated blood to those animals who came back 'empty handed'. This is done without regard to whether or not the bats are related to one another, and the bats certainly do not have the ability to remember the identities of each of the thousands of bats in the roost who may or may not have played fair in the past. This is impressive behaviour, but does it reflect a universal tendency?

This phenomenon is still avidly researched by biologists, but one possible explanation is that the likelihood of sharing food depends on the overall level of cooperation in the group: a 'feel-good factor'. The more bats that cooperate, the more likely it is that even more bats will cooperate. And as the cost to the donor bat is real, but relatively low (they won't starve if they donate), whereas the benefit to the recipient is huge (they might otherwise die), cheating is ultimately counterproductive. Tomorrow, it might be me that needs a donation, so it is in my interest to increase the likelihood that others will help me, by not cheating. Raising the feel-good factor in the colony with a bit of donated blood is a small price to pay for possibly saving my life tomorrow. Such complex interactions show just how powerful a force cooperation can be, and therefore how likely it is to exist, in some form, on other planets.

The last decade or so in animal behaviour research has seen an increasing recognition that animal *personalities* play a very important role in social behaviour, both how animals act alone and in groups. At first dismissed as mere anthropomorphism, it is now clear that different individuals do indeed have different tendencies to each other, and if those tendencies are consistent over time for a particular individual, we call that part of their 'personality'. For example, across a wide range of species from fish to birds to mammals, we see that some individuals in a group are more likely than others to approach an unusual and potentially dangerous object, whereas some may be more aggressive to other members of their group.

Personality isn't restricted to those familiar animals we keep as pets on our rugs. Even zebrafish have been described as having five different personality tendencies: boldness, exploration, activity, aggression and sociability. Anyone observing the birds at the feeder in their garden will notice that some readily hop up to eat, whereas others take their time putting together the courage. This should not surprise us, as we all respond differently to the hormonal and environmental stimuli that we experience, and a large part of those differences is due either to our genetic makeup or to the experiences we went through as juvenile animals. So variation in aggression, for example, is to be expected. But is this variation universal? Quite likely so, because some personality traits in particular – like aggression – are closely linked to fitness and the number of offspring you will have. As with our Hawk–Dove game, on a planet with only non-aggressive individuals a slightly more aggressive animal may have an advantage.

Variation in animal personality has an important consequence

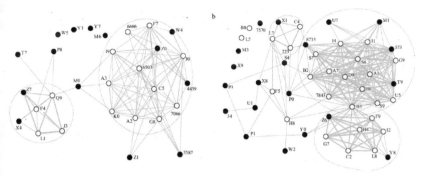

Examples of the kinds of complex social networks that animals must negotiate. These diagrams show the social relationships between individual rock hyraxes (circles, where black is male, white female), with the thickness of the lines indicating how allied any two animals are.* As with humans, some individuals are clearly more popular among their peers than others.

for cooperation. In a group of animals living together, some will be more aggressive than others, and in many cases this arrangement will not be stable. It presents a situation where being the best Hawk is advantageous, but being Hawk #2 is bad news, because you keep fighting and keep losing. In this situation, the aggressive will become more aggressive and the passive more passive. The end result is dominance, where one individual is the big boss, or where each individual has a place in a hierarchy, being dominant to some individuals, but subordinate to others. Dominance hierarchies are an expected result of game theory predictions and are therefore likely to exist on other planets too. But hierarchies bring with them an additional complication: the need for complex brains, so that the animals can *understand* these hierarchies.

* Taken from 'Variance in Centrality within Rock Hyrax Social Networks Predicts Adult Longevity', A. Barocas, A. Ilany, L. Koren, M. Kam and E. Geffen (2011) PLOS ONE 6(7): e22375.

On Earth, we see quite clearly that as cooperative animal societies become more complex, with more individuals and more variation in relatedness, the signals they produce become more complex as well. In a pack of five or six wolves it is relatively easy for each individual to be familiar with the other and to anticipate the response they can expect from a play bow, grooming or stealing a piece of meat. In a troop of fifty chimpanzees, the challenge is much more serious. Greater social complexity leads to a greater need to generate and interpret complex signals, and to remember individuals, their personalities and the history of their past interactions with you and with others.

It does appear that, on Earth, the more complex an animal social structure is, the more complex the cognitive demands on each individual in that society. This shows in the complexity of their communication, in the diversity of their responses to other members of their group, and even in their brain size. Some of the most socially adept animals form alliances, seemingly remembering which individuals have cooperated with them in the past and, amazingly, which individuals have cooperated with *other* individuals, and whether that third party will therefore be a friend or enemy. Dolphins remember acquaintances with whom they have cooperated years or even decades ago, and return to cooperate with the same individuals time after time (often male dolphins, ganging up on females to force them to mate).* Baboons track the social relationships of animals in their troop, knowing which individuals are likely to lend support in a fight and which are not.†

As cooperative societies burgeon, so the demands on each individual to keep track of what is happening in society also become inflated. Similarly, the benefits of reciprocity become

---

* See *Are Dolphins Really Smart? The Mammal Behind the Myth* by Justin Gregg.
† See *Baboon Metaphysics: The Evolution of a Social Mind* by Dorothy Cheney and Robert Seyfarth.

more nuanced: if you can remember who cheated and who didn't, you have a more effective way of deciding when to help and when not. Note that nothing here is specific to the particular biology of life on this planet. A colony of 100 alien 'tribbles' could have just as much selective pressure to evolve complex cognition and alliance building as 100 baboons.

One of the consequences of sociality in animals, taken to its logical conclusion, is that animals will be able to learn from each other. A baby chimpanzee can sit and watch the adults foraging and will be likely to have seen a range of different techniques and skills by the time she becomes an adult. Living in a group exposes you to information, in a way that a solitary existence does not. So we can add to the catalogue of advantages to sociality: social living facilitates information transfer, and for the individual, information is power. Where are the good foraging sites? How can you open this nut? Is this snake dangerous or not?

Passive learning by observation is, as we saw in the last chapter, absolutely fundamental to animal behaviour on any planet, and social groups give you even more opportunity for learning. Those wolves that stay with their parents' pack for longer learn more skills that help them survive when they finally leave. Social groups provide the possibility for *teaching*. For example, a mother cheetah painstakingly teaches her cubs to hunt wounded prey; something that they are unlikely to learn successfully on their own. Adult meerkats bring live (but disabled) scorpions to pups so that they can practise catching them. As social groups become larger and more complex, and brains (or their alien equivalent) become more adept at processing and storing information, the complexity of information that can be taught also increases. Social living opens the possibility of teaching, and also transferring information that can persist in the group for long periods of time: what we would call culture. Faster and more effective information transfer facilitates what we would call technology as well, which could be shaping a stick to retrieve a piece of food, or building a spaceship to come visit the humans on planet Earth.

※

And so to return to the three questions with which we opened this chapter. Why do animals cooperate? Because, as we have seen, they gain a range of benefits, notably protection from predators, and a better ability to find and defend food. The evolutionary process of aggregation can begin for passive reasons – animals can't move away, or don't want to – but cooperation is very likely to continue to evolve from that situation. Under which conditions will animals form cooperative societies? When it is evolutionarily advantageous for them to do so. This could be for reasons of kin selection – if kin selection is operating on the planet in question – or for more abstract game theory considerations. Game theory is a way of predicting how evolutionary advantage will play out, and is a simple technique, applicable on any planet. Kin selection may be vastly different elsewhere in the universe, but if kinship exists in any form – and having some relatives who are more related, and some who are less, will be a phenomenon on every planet – it will drive family-based cooperation as surely as it does on Earth.

And the outcomes and consequences of animals living together? Here is possibly the big surprise of this chapter. There does not seem to be anything particularly parochial about the way that sociality has led to a string of evolutionary consequences here on Earth. Alien animals will follow similar paths: complex social structures, reciprocity, dominance hierarchies, and social learning and teaching becoming more and more sophisticated as societies become larger and more intricate. This appears to be a fundamental feature of the evolution of cooperation, and it gives us very optimistic reasons to believe that, somewhere out there, are creatures that have found a way to manipulate nature in the way that we do. Sociality is so widespread and diverse on Earth, it's tempting simply to assume that it exists on other planets too. Fortunately, our theoretical understanding of why and how sociality is the way it is, gives us a strong indication that aliens will be social too. Teatime with our alien neighbours may be possible after all.

8. Information – A Very Ancient Commodity

We are nearing a goal of some sorts. We understand more about what alien animals must be like: how they move, how they communicate, how intelligent they may be, and also how they might live in groups. We have painted a picture of alien animals that are like Earthling animals in many ways, but they still remain different from *us*, from humans. We are not yet quite ready to answer the question: do they 'talk'? But we cannot put it off for much longer. Whether or not aliens have a language changes everything about who they are and our potential relationship with them, but before we can consider language itself, we have to dig deeper into the nature of communication and, in particular, what kind of information individuals share, i.e. how *much* do they say to each other?

We are constantly reminded that information is a commodity – we live in a world where information has real monetary value. However, we rarely think about how important sharing information is to animals – but it is. It is absolutely vital; no less so than it is to humans with our Facebook and Twitter. For an animal, life or death can turn on the right or wrong piece of information. That's a lot more to worry about than just seeing an embarrassing photo that your mum posts on social media. Precisely because information is so vital, it is subject to strong pressures of natural selection, and the ways that animals acquire and transfer information – reliably or deceptively, as with humans – are diverse and powerful, both here and on other planets. Knowledge is power, in every ecosystem, on every planet.

The information that animals communicate between each

other sometimes shares data about their passive surroundings, but crucially also gives important details about the state and activity of other animals, who aren't necessarily part of the conversation themselves. All animals are active players in a web of information, misinformation and deception. Because of the conflicting interests, and the possibility for animals to misrepresent reality, game theory again plays an important role in understanding how and why animals tell each other anything at all. This puts us in a good position, because conclusions drawn from fundamental game theory considerations are likely to be valid everywhere in the universe. Their worlds may be very different physically, but the mathematical rules of how to beat your opponents in information warfare stay very constant.

But before we can busy ourselves too much with considerations of how 'language' evolved from 'communication' in humans (and why only in humans, and where else it may have evolved), we need to start with a really good definition of *communication*. A really solid definition, because if we make too many assumptions about communication on Earth, we will not be able to apply these ideas to other worlds.

What is communication?

Some things that we do naturally, all day every day, we feel we don't really need to define. We all know what it is to sleep and to wake, to walk and talk, and to ride a bike, even though it's notoriously difficult to describe exactly what you need to do to keep yourself upright while cycling. Definitions are for pedants, some say. But when we start venturing out from our very familiar and consistent world and into unknown alien realms, definitions are crucial. This is certainly the case when we attempt to compare human activities with those of animals – and all the more so when we want to talk about how aliens may behave, and what we might have 'in common'. We cannot afford to accept our routine human behaviours as the basis for labelling how animals or aliens behave. Those parts of our behaviour that seem most obvious to us are the ones most likely to be different in other species.

We all know what communication is in our daily lives, but will that definition suffice for other animals? The *Oxford English Dictionary* defines communication as:

> The transmission or exchange of information, knowledge, or ideas, by means of speech, writing, mechanical or electronic media, etc.

Even if we delete the reference to the medium, we are left with a definition that is too vague for a zoologist: 'The transmission or exchange of information, knowledge, or ideas.' What, actually, are ideas? What does it mean that an animal has 'knowledge'? More importantly, science wants a definition that answers 'why' questions, rather than just 'what' questions. We can see that communication involves the transmission of information. But just observing something tells us very little about the actual nature of a phenomenon. Plus, observations can be very different from different perspectives. Perhaps on a planet where signals travel very slowly – say, slower even than the animals move – the idea of 'exchange of information' may be meaningless, as the situation the animals face may have changed completely by the time the information arrives!

We are far better off with an evolutionary definition of communication, because at least that way we can be confident that our definition is relevant anywhere that evolution occurs. I propose this evolutionary definition of communication:

> A signal produced by one animal and received by another, such that the behaviour of the receiver is altered to increase the fitness of the sender.

Crucial here is the addition of a *mechanism* to this definition, beyond what you might have thought necessary for an everyday definition of communication: *the behaviour of the receiver is altered to increase the fitness of the sender.* In other words, it is not enough that an animal transfers information to another, that information must have an effect. Not just any effect, but this must be an effect to the benefit of the sender. Although that may sound surprising, it's a very universal principle, because certainly nothing can evolve by *reducing* the fitness of the animal that is exhibiting the behaviour. Of course, that's not to say that disadvantageous information transfer doesn't occur: a deer missteps and a twig cracks, alerting a tiger to the deer's presence. But that kind of disadvantageous information can't be the basis for the evolution of a specific trait – in this case, you'd expect the deer to evolve to be more stealthy, not less.

Notice how our new definition remains completely neutral to any specific conditions, other than the presence of natural selection. On our hypothetical planet with extremely slow signal transmission, this definition would still be valid because of the evolutionary mechanism – if the signal is received too late to be of use, it won't benefit the sender and so won't evolve.

You might ask why the communication must be beneficial to the sender, not to the receiver? As long as the sender benefits, our definition is neutral to the effect on the receiver, one way or the other. They may benefit or they may not. In fact, many – possibly even most – communication systems on Earth are mutually beneficial to both the sender and the receiver, particularly when the two animals are members of the same species, and especially so when they are members of the same family. A meerkat signals an alarm alerting others to the presence of a predator, and all members of the group run for cover – the sentry is protecting her family, and thereby her genes. A baby chick chirps with its mouth open wide, indicating to its parent that it wants to be fed – the chick would starve without it, and the mother would lose her investment too. These are mutualistic communication signals that benefit both parties, but mutualism is an optional extra, not actually required for the communication to evolve. The telephone scammer who asks for your bank details definitely does not have your interests at heart, and the cuckoo chick who begs for food from its foster parent is not working for the good of the parent genes (having just thrown the parents' own eggs out of the nest). Benefit may or may not accrue to the receiver, but it certainly must to the sender, so that the behaviour can persist and spread. So, perhaps counterintuitively, we always need to look for the benefit to the sender if we want to understand why communication exists.

Whenever there is a conflict of interest (or a potential conflict of interest) between two animals making their own decisions to maximize their own fitness, then an evolutionary game is in play. Each individual is trying to outmanoeuvre the other – or perhaps cooperate with the other, but for selfish reasons. In the long run, the best gaming strategy is the one that is likely to evolve. So much of how behaviour evolves can be explained using these simple mathematical games, and gaming explanations don't rely on the messy details of the physical world. Do animals fly or crawl, live in the sea or on land? Are they as big as a *T. rex*, or tiny like a flea? Do they swim in warm, saltwater seas, or skate across liquid methane at −200°C? The details matter, of course, but the game is almost identical for all of them. Game theory explanations are therefore particularly good for understanding how life must be on other planets. Think of the universe as being full of little marketing managers of billions of different species. Whatever the medium of communication, or the physical and time scales of interactions, animals on all planets will evolve to optimize their outcomes, and to use information to influence other animals for their benefit too. *That* is the fundamental nature of communication: influencing others for your own benefit.

How does communication evolve?

The first thing that must happen for animals to communicate is that they must be responsive to the environment. This sounds trivial, and it is, because responding to the environment would seem to be fundamental to the very nature of animals (see Chapter Three). But the precise nature of how and to what the animals respond will be different from planet to planet. On Earth, most animals respond in some way to light and to vibration, because both our seas and our atmosphere are transparent (relatively) and also transmit vibrations. Other planets may have physical characteristics that constrain what sensory modalities are effective, or even possible. In a vacuum, sound does not travel. In (most) solids, light does not travel. Even if a signal can be transferred physically, it may not be suitable for communication. Tidally locked planets, with one side always facing their sun in permanent daytime and the other side facing away in permanent night, are likely to have constant hurricane force winds blowing from the hot regions to the cold. Sound seems like a poor choice for a sensory medium in those circumstances. Mostly, we can look at a planet and guess what kind of sensing organs might evolve there.

So can we say for certain that animal communication will evolve only from abilities that evolved to sense the environment? Or is it possible that special communication organs evolve all by themselves, separately from sensing? We can speak and hear because our ancestors detected prey or predators by the vibrations they created in the water around them. But is it possible elsewhere in the universe that, rather than adapting our existing sensory capabilities for communicative purposes, we evolved hearing and speech specifically to be able to communicate? And what about telepathy, a possibility which we mentioned in earlier chapters? Could there be an alien planet with telepathic aliens that evolved this special ability in some other way?

This seems unlikely, although not impossible. If I already have an ear to detect prey, I could also use it to hear signals from other animals of my species – that's not a huge evolutionary leap. But proposing that some communicative abilities evolved all by themselves doesn't provide a good explanation of the step-by-step pathway that evolution must have followed. How did the first communicative organ provide any advantage if, at that stage, no one was communicating? This is a strong argument against science fiction superpowers such as telepathy; we just can't see how it could have got started, if the ability wasn't first providing some other, non-communication, advantage to the animal. That doesn't mean it's impossible, but as part of a scientific enquiry it doesn't yet have a place. We can shed all our assumptions about how aliens communicate, as long as we restrict ourselves to feasible evolutionary possibilities.

Once animals respond to the environment, one of them will start taking advantage of this response, to manipulate the behaviour of others. As stated, communication only evolves if the sender benefits, and there are many, many ways to benefit. The assassin bug hunts spiders by crawling onto their webs and plucking the silk strings to imitate the movement of a trapped insect. The spider, naturally attracted by the sensory inputs that evolved for it to find food, moves towards the assassin bug, which then devours the spider. This kind of aggressive mimicry may not have been the first kind of communication to evolve, but it follows inexorably from the fact that manipulating the responses of other animals can most definitely be to my advantage. One way or another, alien creatures will be exploiting the gullibility of others.

After such simple communication comes to exist, then both senders and receivers will refine their signals and the way that they respond to them, to build a richer suite of behaviours to serve them better. Senders adjust their signals to be clearer: perhaps louder, more prominent, more distinct from other sources of noise in the environment. Birds preferentially use particular song notes that travel well in their particular environment: tweets work especially well in woodlands, and warbles in grasslands. The receivers, too, refine their responses, perhaps responding only to the songs of their particular species, or evolving a complex system of duetting, where signals are sent back and forth so that each member of the conversation can decide whether or not they trust the other (perhaps to be their mate). This is what drives all the complexities and diversities of animal communication that we see on Earth – just looking around us, it is clear that animals are not merely using the sensory proclivities of other animals to send deceptive messages. Animals have evolved entirely new and unique communication systems that are dedicated to conveying particular messages. Birdsong is not a phenomenon that existed before birds, rather, birds have evolved both complex ways of producing song and complex ways of interpreting what it means, on the back of their innate ability to perceive sound.

Even though this diversity of animal communication results from common evolutionary mechanisms that must be shared by creatures on other planets (the exploitation of existing sensory cues to create manipulative messages), the details are utterly specific to our own planet. We can be confident that alien animals will communicate with each other for all the same reasons as they do on Earth, but what the precise nature of that communication will be – what their equivalent of birdsong is really like – depends on the physical conditions on that planet, such as the thickness and composition of the atmosphere. These physical constraints were a big feature of Chapter Five, but in this chapter we need to talk less about *how* aliens communicate, but rather, *why* they do.

Decisions, decisions

Communication evolves because animals make decisions. Decisions about where to go, what to do, what to eat and (on Earth at least) with whom to mate. Communicative information influences those decisions – that is why communication evolves. Sometimes that influence is exploitative – for the benefit of the signal sender and to the detriment of the receiver – and sometimes it is mutually beneficial. Selfishness and altruism go hand in hand in animal communication. We are apt to think that, as deception is so common and widespread in humans (not just in human politicians), it is a uniquely human characteristic. Similarly, our most uplifting altruistic messages imploring compassion and empathy seem to define the best of humanity.

Both truth and deceit are, of course, amply present in the animal world, and will be so on alien worlds too. In fact, truth and deceit are less different than you might think. From an evolutionary perspective, both are, after all, simply different ways of increasing your fitness. Darius the Great, the king of ancient Persia, realized this in 522 BCE when he said:

> Those lie whenever they are likely to gain anything by persuading with their lies, and those tell the truth in order that they may bring themselves gain by the truth.★

Decision-making is liable to exploitation, and this is a universal feature of all decision-making because there is no such thing as the perfect decision. Perfection is never achieved, because every benefit carries a cost, and both cost and benefit must be

★ The ancient Greek historian Herodotus reported this as Darius' speech to his co-conspirators while plotting to seize the throne from a usurper (*The Histories* III, 72).

traded off against each other. This is a fundamental principle of evolution, and decision-making is no different. Perhaps I should always wait until I have perfect information before making a decision – I will invest in a company, but only when I am 100 per cent sure that it will succeed. A spider could decide only to approach a struggling object in its web if it is 100 per cent sure that it is not actually an assassin bug. Such strategies are unlikely to be optimal, and are in fact going to be disastrous, because 100 per cent certainty can never be achieved! Information is incomplete, error-prone and even downright deceptive. The trick is to use information optimally, and to take into account the costs and benefits of each decision, including the chance of error. A balanced decision – cautious, but not over-cautious – works best in your evolutionary favour.

Communication broadens our access to information and, we hope, makes our decisions better. For the signaller, communication provides us with the possibility to influence the decisions of others; how we dress and how we behave influences whether others will hire us, fire us, sleep with us or beat us up for supporting the wrong football team. So our communication is dependent on influencing the decisions of others.

Exploitation, of course, is not everything. Often, animals will communicate so that they can cooperate – humans do this all the time. But cooperative communication lays you even more open to exploitation, even if on the face of it cooperation is what the communication is all about. We saw this in the previous chapter with the Prisoner's Dilemma: cooperation is great, but that doesn't mean it's going to be preferred in the long term. As the receiver of information, I need to decide how reliable the signal is: am I being deceived? If I take all signals at face value, assuming that the sender has my interests at heart as well as theirs, I am certainly open to exploitation. Inevitably (evolutionarily), a cheater will arise and take advantage of my gullibility. Naturally, I can neither fully trust nor fully distrust every signal, so my decision is a trade-off. Can we find any circumstances under which I should be more trusting rather than less – circumstances that will apply on alien planets as well as on Earth?

Evolutionary game theory shows us two universal reasons for me to be trusting – and these are likely to have similar effects on animal communication on other worlds too. The first is if the sender and I share common interests. On Earth, this is usually because we are related; kin selection should drive the sender to be cautious about harming my interests – they share some of those interests. I wouldn't lie to harm my kids! If aliens have kids, they will feel the same way. But there are other forms of shared interest besides relatedness.

If my interlocutor and I are engaged in a critical task where we would both suffer should the task fail, we have a shared interest. Cooperative hunting is the most familiar example of this, because no one of the individuals taking part could achieve their goal (bringing down an elk, say) on their own. A cheating hyena that doesn't take part in the hunt is obvious to the others, and those animals who put in all the effort can simply deny the food to the freeloaders. Kin selection can explain part of cooperative hunting (wolves in a pack are generally related to each other, although packs do often take in unrelated individuals), but cannot be the only explanation for this cooperation. Alliances of male dolphins, for example, are less closely related, and instances of pods of fifty or more orca attacking and killing blue whales probably consisted of unrelated individuals. In a situation where everyone has to pull together or no one benefits, signals are more likely to be reliable than manipulative, and this is true on any planet, whatever the kinship relationships between the individuals. If we ever discover signs of cooperative ventures on another planet – perhaps city-like constructions, or even space-going probes – our first conclusion will not be about alien intelligence or alien technology, it will be that these aliens *communicate* to cooperate.

The other reason for signals to be reliable is if they are very costly to produce. In the age of fake news it is easy and simple to spread unreliable information such as 'the Earth is flat', or 'vaccines cause autism'. Lies are cheap. However, expensive messages are hard to fake. When you see a giant pyramid that says 'Pharaoh Khufu was a powerful emperor', we tend to believe the message; how could Khufu have built that pyramid without *actually* having an army of slaves? Similarly, many animals send signals that are simply too expensive to be unreliable: the massive antlers of red deer, or the bright colours and long feathers of a peacock. If those individuals are faking, and aren't really fit and healthy individuals after all, how could they have produced such messages? It's a universal principle behind the evolution of signalling that the fitness cost of a signal, and the shared interest of the sender and receiver, will determine the perceived reliability of that signal even under radically different conditions on another planet. This is a deeply fundamental claim we can make about alien communication: reliable alien animal signals will be costly or will reflect a fundamental shared interest.

So how many different kinds of information are there among the animals on Earth? And do we think that this diversity represents *all* of the different kinds of signals that could exist anywhere in the universe? There seem to be at least three 'topics' of information that will be important to animals on every planet: information about the environment, information about the individual signaller, and information about group relationships. Let us look at each of these separately.

Information about the environment

What do animals on other planets have to tell each other about the world in which they live? We can be confident that there are some interesting features of the environment that will be present on all planets: all life needs food, and needs to avoid predators, and so we can learn about alien food and danger signals from those we have on Earth. But perhaps other planets also possess particular dangers or opportunities that we never experience on Earth, and the animals there must communicate with each other about environmental conditions for which we have no conception. This is certainly possible, and might be difficult for us to understand. Perhaps on some planets individuals need to warn each other about impending magnetic-field distortions – something that would go quite unnoticed by most species on Earth. Some aspects of alien communication will always be alien to our comprehension – even if we can decode its meaning.

However, while we can't claim to understand *everything* about alien communication, we can confidently say that all life needs energy, and so food will be something important to talk about on any planet. So we shall begin by looking at information on food and predators.

If you discover a particularly delicious-looking strawberry patch hidden in the woods, would you call everyone's attention to it, so that they could come and share? Maybe. Certainly, you might call to your family; that's kin selection again driving the communication of information about food. But apart from relatedness (as we don't know what alien relatedness is like), is there any reason to give others information about your newly found food source? Actually, yes, there are other reasons why an animal might signal the presence of food to attract others to come and feed. For one thing, it is safer to feed in a group – if your friends are around you, it is less likely that a predator will be able to sneak up on you.

True, that means less food for you, but this is better than becoming someone else's meal. Small birds like chickadees feed in flocks that include other species like nuthatches and titmice – each species has its own specialist eyes and ears, and it's worth sharing your food for the extra protection this provides from predators.

When feeding in a group, you are also better protected from other, possibly larger animals that are interested in the same food source. Ravens make loud calls when they discover a food source that is difficult to access, like a carcass being scavenged by other animals. The calls attract other ravens, and a large flock of these birds is a significant deterrent to other scavengers. Friends can help you get hold of the food, keep hold of the food, and keep you safe while eating.

For those animals with more complex foraging patterns, sharing food information takes on additional complexity as well, because they can talk *about* the food. Chimpanzees have distinct calls that they give for different types of particularly tasty food ('mango!' is different from 'banana!'), but they use a much more general call for a food discovery if it is only mildly tasty ('it's just carrots or apples or whatever . . .'). An important signal ('tasty food!') needs to be clear and specific, much more so than a signal of lesser importance ('boring food'). True, we can't be sure that an alien planet will have food that varies so much in appeal (perhaps only one kind of fruit grows there), but food will always be a limiting factor for any ecosystem, and at the very least the quantity of food present will be an interesting topic for conversation.

Messages are also shaped by the temporal context of the information they contain. If an animal needs to make a long-term decision, for example which mate to choose, then a signal that doesn't change much over time would suffice. A bird's plumage reflects the fitness of the male, and even if the colours fade somewhat as the animal ages, the timescale of the signal matches the timescale of the decision that needs to be made. On the other hand, a predator alarm call needs to be instant. The decision to flee or not needs to be taken immediately! If predator alarm calls were a long, rambling explanation of the nature of a predator approach and the dangers it poses, the predator would already be upon you before you'd finished decoding the message, and that doesn't make for an evolutionarily effective signal.

Indeed, animal alarm calls are typically short, clear and grab the attention of other members of the group as effectively as possible. There is also evidence that alarm calls which indicate a disturbing danger should be, in themselves, 'disturbing'. They grab the attention of other animals by their very nature. On Earth, where most alarm signals are given by sound, it does seem to be the case that many alarm calls have acoustic properties that are, in fact, inherently alarming – called the 'sound of fear'.★ Screeches and screams are disturbing to humans, but also to other animals. Female deer react to the sound of a human baby crying in much the same way as they do to baby deer alarm calls, and the sound of an animal in distress is as disturbing to us as a human cry.

There may be a general evolutionary principle at work here, applicable to aliens too. The kinds of sounds that we – and mother deer – find disturbing have certain acoustic properties. The adjectives we use to describe such sounds, like 'screechy', 'raw', 'coarse', refer to the unpredictable way that the sound frequency varies. Alarm calls tend to consist of chaotic, almost random variations in sound frequency, and that is fundamentally an upsetting thing to hear. Now, one possibility is that these sounds are disturbing because of that very unpredictability – a reasonable hypothesis, as unpredictability is not a nice thing for any animal. But also, these chaotic frequency variations are

★ For more details on this, I recommend Dan Blumstein's TED× UCLA talk on 'The Sound of Fear': https://tedx.ucla.edu/talks/dan_blumstein_the_sound_of_fear/.

simply a mechanical result of making sudden and extreme noises – when you're panicking and you scream, your vocal cords vibrate in a disorderly way. It's similar to the distortion you get when you turn the volume on your amplifier all the way up. Pushing any sound production mechanism beyond its usual limits makes chaotic noise. So if alien animals use sound for their alarm calls, *their screams will probably be very much like ours.* Don't believe it if they say 'no one can hear you scream' – screams evolved to be heard, and to be disturbing. Even if aliens don't use sound, it's likely that alien alarm calls will be similarly chaotic in whatever medium they do use. They will have whatever properties are characteristic of the alien signal-production organ when you jump out from behind a rock and give the alien a fright. 'Scary' is going to be similar on every planet.

To illustrate one way this could work, consider Earthling fireflies that flash a carefully regulated pattern using the light-generating organ in the male's abdomen. Each species has a particular timing for its flashing pattern that avoids the embarrassing confusion that would happen if a female should be attracted to a male of the wrong species. When hundreds of fireflies of the same species congregate in the same area (notably in the Smoky Mountains in Tennessee in the early summer), a spectacular phenomenon can occur, in which the individuals synchronize their flashes so that they all pulse on and off at the same time. Now, imagine some kind of similar alien visual communication system.* If one of our alien fireflies, currently engaged in synchronized flashing, were to spot a predator, it might suddenly change its flashing pattern, which would break its synchrony with nearby creatures. The resulting asynchrony could cascade through the network of fireflies, creating a

---

* Although it appears that fireflies on Earth do not use their flashes to convey complex information in this way.

suddenly chaotic flashing of lights that would be a visual equivalent of the sound of fear: a 'sight of fear'. Just as we get a headache listening to babies screaming, the alien fireflies would probably find this chaotic flashing equally disturbing.

Information about the individual

Given that communication between at least two individual animals is both exploitative and cooperative, it makes sense that animals want to know as much as possible about their communication partner. That could be as simple as the identity of their co-correspondent (assuming that the animal has a memory for individuals), or some important details about them: where they are, how big they are, how angry they might be . . . It's always best to know as much as you can about someone before you open your mouth.

Some information about an individual is inherently reliable. I can show how big I am by standing on my hind legs. Of course, I might want to exaggerate my size, by puffing up my fur like an angry cat, but there's a limit to how much I can get away with: I am only as big as I am. Red deer can tell from the roars of a male how big he is, because much like an organ pipe, the depth of the pitch depends on the size of the animal. Although male deer have evolved the ability to lengthen their own vocal tract to roar with a deeper voice (and so sound bigger than they actually are), when everyone is faking, big deer still have a deeper roar than small deer. Individual identity, too, is rarely faked. Many animals – including humans – can recognize their mates and their offspring by the subtle variations in their appearance and in the sound of their voice. King penguins must be able to identify their chicks, and chicks their parents, in a colony of thousands of animals, and they do so by the specific individual differences in their vocal calls. Individual identity is remarkably widespread.

Of course, wherever a signal is assumed to be reliable, cheaters will try to exploit that reliability, such as cuckoos masquerading as the chick of their hosts. In this case, the hapless parents have a simple strategy to get them out of the dilemma: ignore all information and feed the chicks anyway. Information is not always the solution. The dance that animals must play between relying on a signal and distrusting it is dizzying and never-ending. That trade-off is not an Earthly phenomenon. All alien animals will evolve to be sceptical – balancing the gain of trusting information against the risk that you are being deceived. The fact that on Earth we see the whole range of strategies, from the 'trusting' reed warbler who raises a cuckoo chick to the penguin who feeds his chick and his chick alone, is strong evidence that on other planets there will be the equivalent of naïve warblers and discerning penguins too.

Unreliable signalling about an individual is most likely to evolve where it bears little cost or, rather, where that cost is manageable. Just as costly messages tend to be reliable, lies are cheap. But it's also possible to lie when you have the option of backing down and changing your message at short notice. We see that around us in the ritualized displays of so many animals on Earth. In many species of birds, for instance, copying the song of a rival is a red flag, a clear aggressive statement. It's similar to what would happen if you were to pick someone in a pub and start copying everything they say (and copying their accent as well). Pretty certain to get you into a fight. But this behaviour is crucially different from genuinely costly investments, like growing large antlers, or colourful feathers, or going to the gym every day to build your muscles. Annoying someone in a pub is a cost you can choose to bear, or choose not to bear. You can *manage* the cost. A bird can decide to copy the song of a rival today, or not today but tomorrow. And if the rival looks just a bit too hard to handle, you can always retreat. Bluffing, a form of deceptive signalling, is risky, but sometimes worth the risk. Alien animals would almost certainly perform ritualized posturing (a kind of communication), precisely because it is an almost inevitable consequence when a signal has manageable cost. That's not to say that aliens will definitely play poker, but like on Earth, alien animals will definitely have to take some very similar decisions on when to gamble, and when to fold.

Information about my group

In a group, there's so much more to talk about. Communication reaches its zenith in group dynamics, and the complexity of what you want, and need, to say spirals upwards together with your cognitive ability to think about things in a more complex way. Where groups are in conflict with each other (which they always are, even if they're not in openly violent conflict) there must be a way to tell one group from another. Sometimes the advantages of distinguishing your group from another are clear; most likely you have developed ways of cooperating that keep the group functioning well. Your long-term relationships with other group members are known to you and to others so that your social structure remains stable.

If a group of animals have a dominance hierarchy – or even if they live in a more egalitarian arrangement – then if everyone in the group knows the rules, the social wheels turn smoothly. Many highly social species, hyenas and chimpanzees for example, have clear dominance relationships, and overstepping your mark can have some pretty aggressive consequences. Other species, hyraxes for instance, have a more egalitarian society – and this suits their niche very well. Either way, the social structure is crucial to the survival of the group. Introducing a new individual into a group disrupts the dynamics; by knowing in which group you belong, and not joining the wrong group, you can avoid inadvertent conflict.

Other advantages to signalling group identity are less obvious. Many songbirds aggressively maintain their own territories, but respect the territories of the males immediately surrounding them. They learn to recognize the songs of their neighbours, and do not consider them to be a territorial threat. If this were human politics, we might call it 'detente'. When a male stranger arrives, however, he is recognized as an outsider to this group – even though the collection of birds, each in their own territory, is not actually a cooperative group in itself. As a result, he receives aggressive responses from everyone – this is known as the 'dear enemy effect', because neighbouring males are 'enemies' in theory, but ones with whom you can get along. Information about social relationships within and between groups tends to stabilize those relationships. Territoriality is fundamental when resources are limiting, so alien planets are likely to have the alien equivalent of territorial songbirds. What's more, those alien songbirds should advertise and react to messages according to the dear enemy effect too.

Overall, signalling group identity is vital for enabling an animal to decide whether or not to trust another individual. Dolphins live in groups that often break apart and come together again, over and over. Having built a social relationship with certain individuals, it is important that when the dolphins reunite, they can recognize the group with whom they cooperate well and the individuals of which they are more likely to trust. This is particularly true when animals don't just need to get along peacefully, but actually need to coordinate their activities. Dolphins do this by referring to each other using special whistles that act as individual names. Each dolphin has a very characteristic whistle called a signature whistle that is used as an individual identifier, both by the dolphin to whom it belongs and by other dolphins wanting to address it. Remarkably, dolphins can remember and recognize the signature whistles of individuals even after being absent for decades.

But if the group is large, recognizing everyone's identity places large demands on an animal's brain. Some animals (notably humans and other primates, as well as dolphins) have evolved very large brains, quite possibly in large part to deal with keeping track of their 'friends list'. An alternative 'low tech' solution is if everyone in a group signals their membership of that particular group, so that it's not necessary to remember everyone's individual name. Wasps can recognize hive-mates by the smell of the nest, which rubs off on any individual living there; there's no need for individual recognition – if you smell like home, you belong at home. It's rather like after a football match, when you can instantly spot which pub is full of those wearing blue, or full of those wearing red, and you draw your conclusions about what would happen if you go into either pub.

But of course, humans do much more in groups than merely drink and sing. We build cathedrals and radio telescopes and spaceships, and find ways to understand aliens, hoping one day to meet them. We *coordinate* our group activity, and to do that we need to communicate very complex information, from mathematical formulae to design plans to book manuscripts. It may be surprising that very, very few animal groups show that kind of coordination. There simply aren't many examples on Earth of animals that communicate enough information to coordinate in complex ways.

Social insects may, at first glance, be the exception. In Chapter Six we talked about how honeybees reach consensus decisions about hive relocation, and how ants and termites also achieve spectacular construction feats that presumably could not be successful without some form of coordination between thousands of individuals working to a common goal. But this coordination is different in its nature from what we humans think of as coordination: the construction of the Cambridge University Mullard Radio Astronomy Observatory, for example. Consider those annoying ants that always seem to be able to find a way into your kitchen; in regimented lines they come, following a slightly winding path that only varies when you place some obstacle in the way. They are coordinated, yes. But their coordination is an emergent property of a very simple process – apparently complex behaviour arising from very simple behavioural decisions. They are smelling their way via the footprints of hundreds of ants who passed previously. They have one single coordination signal: 'I was here.' True, this complex behaviour also relies on the fact that as the smell evaporates, the ants can tell how long ago their colleagues visited that spot (otherwise the kitchen floor would be a mess of conflicting signals). But the signal, though it varies in strength, is just a single piece of information. Finding food (or enemies) is about all the ants can do with it.

Now consider coordinated hunting by Ice Age humans. To bring down a mammoth, they must each understand their role in the venture and respond to the movements not just of their quarry but also of the other members of their group. Each individual must respond to the responses of others, and hand and vocal gestures probably evolved to help an individual figure out what the others wanted him or her to do. This is a very complex coordination task and, it seems, beyond the ability of almost all animals on Earth. The information you need is rich, fast moving and sometimes contradictory. But this is the kind of communication that an alien species would need to evolve if it's going to coordinate building a spaceship to come visit us.

There is so far some limited evidence that chimpanzees can take up specific roles in a hunt, and evidence that is more anecdotal still that orca use vocal signals to coordinate their efforts to create large waves and wash seals off ice floes. But unless we can understand more about how animals signal complex information to each other, these examples of using communication to coordinate their activity are very rare. In general, just one species on this planet – us – can coordinate in this way on a regular basis. Coordination may be rare on other planets, or alien worlds may be swarming with organisms all performing detailed coordinated tasks with each other. But on Earth, this is an exceptional behaviour.

Incredibly, though, this lack of coordination in animals has nothing to do with the amount of information that they *could* transfer. Most songbirds, as well as dolphins, wolves and many other intelligent species, have communication systems that could potentially contain far, far more information than would be needed to coordinate activity. It's like they are constantly chattering away, but without saying anything. Information is missing. The available bandwidth is filled with static. Why?

Information content: how much *can you say?*

The starling is a very noisy and well known bird in much of Europe that sings long and complex songs. Male starlings can produce tens of different kinds of 'notes': some whistle-like, some warble-like, some rising in pitch, some falling, etc. They tend to group notes together into song 'motifs' that a particular male uses repeatedly, although motifs are different between males. The number of different ways that these motifs could be arranged (let alone the notes within the motifs) is astronomical. Blackbirds sing equally complex songs that are rarely the same twice. The American mockingbird uses notes drawn from a repertoire of up to a hundred different song types. Each of these birds undoubtedly has a sufficient repertoire to recite all of Shakespeare. The information *potential* is there. But birds

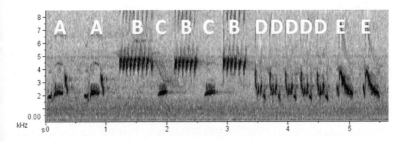

A sequence of mockingbird song showing how different call-notes are repeated and combined. This particular song has fourteen call-notes of five different types. If the notes could be arranged in any possible order, there would be over 6 billion different combinations!

still only use a handful of different motifs – a tiny fraction of the possible combinations of motifs available to them. Why would they have such huge information potential, and not use it?

It's the evolution of information potential, and what use is made of it, which determines how, and whether, language would evolve on another planet. So many species on Earth, from starlings to hyraxes to humpback whales, communicate in sequences of discrete sounds of different types – notes, in the case of birds. So do we, of course. English has about forty different basic sounds called phonemes, vowels and consonants, which can be combined into all the different words in the language. If we limit ourselves to words of five phonemes or less, that's still more than 100 million possible words★ – many, many more than actually exist. Why is there so much diversity, more than even we as humans need? Having more information capacity than you actually need seems to be something we have in common with the notes of starlings and the groans of whales.

★ The number of different five-phoneme combinations (with forty phonemes from which to choose) is $40^5 = 102,400,000$.

Of course, a language can only evolve based on a communication system that can support a very large quantity of information – if not an infinite quantity. An alien planet where none of the available communication modalities (see Chapter Five) could support such a large variety of signals could not support any species with language. But it's still a puzzle that Earth contains thousands of species that communicate using sequences of different sounds – essentially equivalent to the phonemes we use for speech – and yet only one of those species has a language. Just having the information capacity isn't enough to mean that you use it, and it's a little perplexing that natural selection should be so wasteful in giving starlings so many notes that they don't need.

As we have often found in this book, it is the theoretical consideration of why a particular behaviour evolves that seems to give us the most general and universal insights into what might be happening on other planets. Mathematical models can show the trade-off between being able to share extra information, and the cost of developing and maintaining a complex brain that can both produce and interpret these signals. Naturally, there is likely to be an optimal balance. Individual male starlings want to distinguish themselves from other males, so they need a certain amount of complexity in their song. Probably the cheapest way to do this (for the brain) is by producing a sequence of discrete sounds. A sequence of notes is distinctive, catchy and easily recognized.

In contrast, the alternative would be if each male tried to distinguish himself only by some continuously varying feature of his song – song length for example. Then they would need immensely complex ears and brains to distinguish between more than a handful of different individuals. Imagine having to tell the difference between two birds on the basis of one song being just 0.1 seconds longer than the other! Complex song sequences of discrete notes are actually pretty easy to produce and recognize, making them the most evolutionarily efficient way to send messages. Songs made up of discrete notes may have the potential to recite Shakespeare, but that's not why they evolved – they probably evolved purely because they do a simpler job more easily, and the extra complexity potential comes for free. The evolutionary phenomenon whereby a trait that evolved to provide a particular advantage is reused as the basis for a different faculty is remarkably common. Bird feathers were probably used to keep their dinosaur ancestors warm, but turned out to be particularly useful for flight. Diverse sequences of sounds make it easy to identify individuals, but can be harnessed for the full force of language itself.

But this raises two vital questions. Firstly, is it inevitable that *if* a planet lends itself to complex modalities like our acoustic modality (with birdsong being a good example of its complexity), then sooner or later some species will evolve to use that modality as language? Once the potential is there, will evolution eventually take advantage of it?

Secondly, is it *only* via these sequences of discrete elements like phonemes or birdsong notes that sufficient information could be assembled to produce a way of communicating that we would call 'language'? Or perhaps there is some other alien way of communicating huge amounts of information without phonemes?

These questions lie at the heart of what alien civilizations will be like, because it seems absolutely inevitable that such civilizations must use a language of some sort. To the first question – is language inevitable should a planet lend itself to complex modalities – I believe that the answer is yes. The difficulty is in the timing; 'sooner or later' can have different mathematical interpretations.★ But to the second question – are sequences of discrete

---

★ The question of how long a difficult task like evolving intelligence will take is discussed further in *The Cosmic Zoo* by Dirk Schulze-Makuch and

elements the only way this could happen – there is good reason to think that the answer is no.

We know that some highly intelligent species on Earth do *not* use sequences of discrete sounds like phonemes to communicate. Sound sequences are only one way to compose complex messages, which works for some species like birds, but not for all. Dolphins, in particular, are generally considered to have a highly complex acoustic communication system. Certainly, as discussed earlier in the chapter, they are the only species we know (apart from humans) in which individuals mention each other by name: their signature whistles. But crucially, this whistle is not a sequence of discrete sounds like the names that we use – for example, 'ch-ar-uls', 'dar-win' – rather, dolphin signature whistles are a continuous up–down sweep of pitch, rather like a police siren. Nonetheless, each signature whistle is different, and that means they could form the basis of more complex communication as well.

I said a few lines ago that a starling would have a hard time hoping to identify individuals by some continuously graded property of their song that can take on any value, like song length or pitch. Yet dolphins do this all the time; as do wolves – howls are another good example of a complex signal not based on a sequence of discrete elements. The reason why two very intelligent and very social animals like wolves and dolphins have bucked the obvious trend of using sound sequences to communicate simply has not been investigated enough yet, and this is one of my own main fields of research. By observing the way that the animals behave in the wild, and at the same time analysing the sounds they make, we can ask: do two different howls

---

William Bains. Essentially the distinction is between a 'random walk' model, where the outcome is not guaranteed to occur in any arbitrary long period of time, and a 'many paths' model, where the outcome is likely to occur in any fixed period of time.

really have two different meanings? Perhaps one to say, 'come here', and another to say 'go away'? Or perhaps that is an overly anthropocentric way of viewing information – what *we* expect to read from a message, rather than what the animal needs.

However, it's a good guess that the answer to why wolves and dolphins use such a simple information channel lies in the physical properties of the environment over which the animals are communicating. Wolf howls and dolphin whistles are used to communicate over long distances where sound becomes weak and distorted. If a message depended on a precise sequence of elements, there's a good chance some of them would get lost in transmission, and the whole message would become corrupted. Continuously varying whistles and howls may be a way around this distortion problem. On a planet where the communication channels (acoustic, visual or otherwise) are very noisy – perhaps howling winds, or a dusty, cluttered atmosphere – information can only be reliably transferred using these simpler, slower methods, like howls and whistles.

But could continuously varying signals – not sequences of discrete sounds – eventually evolve into language? Most human linguists think not,[*] but my opinion is that we don't yet know enough about dolphin or wolf communication to make such a broad claim. It's worth remembering, though, that the distinction between discrete words and continuous whistles isn't as clear-cut as it seems at first. My children always insisted that the phrase 'It's a dog-eat-dog world' is actually 'It's a doggy-dog world', and that the 2018 Wes Anderson film *Isle of Dogs* is actually 'I Love Dogs'.[†] The gaps between words that we see on the printed page aren't

---

[*] Steven Pinker makes this point very forcefully in his popular book *The Language Instinct: How the Mind Creates Language*, although not everyone feels there is enough evidence to support this.

[†] These are known as 'oronyms'.

always present in speech. In some ways, we actually do speak in continuous sounds, rather than in birdsong-like notes.

Animals evolve a particular communication channel largely because of the physical constraints of their communication: are they speaking across wide valleys or narrow cracks in the sea floor, or through sludgy, half-frozen benzene on an ice planet? But information requirements are determined by the social needs of an animal, not its physical environment. If you need to talk, talk! Evolution will adapt whatever existing communication medium is practical. So despite the apparent paucity of information in continuous signals, we shouldn't exclude the possibility that aliens can, in fact, use such a strategy. If complex communication using continuous signals has evolved in multiple species on Earth (albeit without becoming a true language), there's a good chance that on other planets, where the environmental conditions favour such a way of representing information, the evolutionary pressures could well lead to whistling and howling becoming the dominant ways to communicate complex messages. The possibility is there, and there seems no evolutionary reason why whistles should not form the basis of a language.

Overall, then, our examination of how animals exchange information has shown us quite a few ways that we can expect to be able to predict how aliens will behave. Most likely aliens will send messages through the physical channels for which they have evolved their sensory organs. Their communication will, like on Earth, be fundamentally selfish, evolved to benefit the sender first, and with cooperative, mutualistic communication arising only later and under specific conditions.

We can make some general predictions based on game theory about what kind of messages they may send to each other, but the precise details of their birdsong or firefly flashes will depend on the specifics of each planet's conditions – do they use sound or light, or even electric fields? However, as we saw in Chapter Five, the communication channels that are effective on any particular planet are relatively constrained by the physics of the environment. If we can say that an alien will use sound to communicate, we will know to look for information in the combination of frequencies in that sound. The process of decoding alien communication will be similar to the way that we go about decoding animal communication on Earth. If we know where in a signal the information is contained, and we know how that information changes in different situations, we can start to infer the meaning of the message.

If individuals are 'related' to each other, they will be more likely to trust messages from their family members, and if not, we can still expect 'costly' messages to be more reliable. Alien deer may well have large antlers. They will talk to each other about food and about predators, and it's even likely that their predator messages will be 'scary' – although perhaps only scary to them, if they use a very different communication modality to us. Most likely they will identify each other using signals, and quite probably they will be able to identify groups, if they do in fact live in groups.

Finally, for those aliens that go on to develop a language, as we hope they will, so that we can talk to them, their communication will be built upon a structure that can support an essentially infinite amount of information. On Earth, this is mostly done with sequences of discrete elements, and this is probably commonplace throughout the universe. However, we shouldn't close our minds to the possibility that other information-carrying techniques are possible, and our language as we know it is almost certainly not the only way to form a language. This final point is the topic of the next chapter: what exactly must a language be?

9.   Language – The Unique Skill

'They are one people, and they all have the same language,
and this is what they start to do? Now nothing that they
want to achieve will be impossible for them.'

GOD, Genesis 11:6 (The Tower of Babel)

Language seems to be the single thing that makes us unique among the creatures of this planet. Every other supposedly unique feature that humans possess, animals possess too: tools, culture, emotions, planning, even humour. It is only language that we haven't been able to find anywhere else on Earth. Language gives us a unique ability: to see into the minds of others in a way that we will never be able to see into the minds of animals. Language also shapes the way we think and makes us who we are. It drives and enables our cooperation, and so is responsible for the greatest achievements of humanity. But we still don't really know what it is.

Surprisingly, it's very hard to define language, although we'd like to be able to, so that we can distinguish ourselves clearly from other animals. We can't even point to language's set of unique characteristics – if it has any, that is. And that's a shame, because if language had clear-cut characteristics then we could recognize language in aliens when we meet them. But we don't even know whether language is an ability that a particular species either has or doesn't have – or conversely whether there is a gradation between some species that have more language than others – perhaps dolphins or chimpanzees. In fact, we don't even know if language is necessarily a single 'thing', if it is one kind of fundamental structure that must be shared by every 'talking' civilization across the universe. Alternatively, language may be just an ability, a functionality, which could be implemented in any one of many different ways.

For something that's so important, we don't know much about language. Perhaps that's not surprising, and plenty of science fiction authors – and scientists – have toyed with the idea that alien language is likely to be not just unfamiliar, but inherently incomprehensible – so fundamentally different in nature that we could neither recognize nor decode it. We are going to attempt to address these dilemmas first by asking what language is on Earth, and then seeing whether this is how language must be everywhere. Once we are clear on the nature of language, we will examine the different ways in which it might evolve.

There are quite a number of scientists (linguists, for example) who would take issue with my claim that we don't know much about language. We do, in fact, know rather a lot about *human* languages as a whole. We know what they're made of, how they are assembled and what the different parts do. And we know how different languages use these different structures, like nouns and verbs, clauses and basic sounds like phonemes, often somewhat differently from language to language. But human languages really are very, very similar to each other – although this is not much consolation when you get as frustrated as I do trying to learn an unfamiliar language later in life.

But our languages really *are* similar. Even the concept of combining a verb (a 'doing' word) and a noun (a 'thing' word) is something that doesn't *have* to be shared between languages, but is. Scientists like Noam Chomsky have claimed that, more than just sharing components (like nouns and verbs), every language actually combines them together in a very similar way, following well-defined rules that can even be represented mathematically. That doesn't make it *easy* to understand Amharic if you don't already speak it, but it might help us understand aliens, if we put our best minds and computers to work at it.

Now, this throws up our first important (and as yet unanswered) dilemma. Is language – and I mean language applied across the universe, not just on Earth – something that is defined by some set of mathematical relationships that we can write down, a set of rules that must always be followed by anything that claims to be 'language'? Or is language an ability, something that we should define functionally, as in, 'that which allows me to communicate a complex concept to you'? If there is in fact one universal rule (or set of rules) that every language must obey, then we are in luck. We will indeed know a great deal about alien languages, because we should be able to derive those rules from looking at human languages. But if there are no such fundamental structures, then language is just 'talking', and alien language could be permanently unintelligible to us, sharing nothing in common with what we know about communication. So does human language have a set of fundamental rules? Are those rules shared with animal communication? And is there any reason to think that they may be shared also with aliens?

Infinity from the finite

One of the fundamental assumptions linguists make when deriving the rules of language is that language must be infinite. We are happy to understand that our language is infinite, because we feel intuitively that there will never be an end to the books that could possibly be written, or to the number of unique ideas that you could express. The infinite nature of language is taken for granted – almost taken as *the* distinction between language and 'just communication'. As far as we know, no other animals on Earth can communicate more than just a handful of different concepts. So we're pretty unique in being able to write books, as well as poetry, lyrics and political speeches.

But how is this infinity achieved? We only have a handful of words. Well, we actually have quite a few words. English has, at a very conservative estimate, at least 170,000.* That sounds like a lot, but it's less than the number of books published in the UK each year, so clearly we need to represent a far vaster array of concepts than we have words available. Noam Chomsky pointed this out in the 1960s and claimed that the fundamental difference between language and non-language is that language is something beyond the trivial combination of words into longer and longer sentences. No animal, or even alien, has a brain big enough to process a language where an infinite number of concepts is represented by infinitely long sentences. This is, in fact, the core of the language dilemma: there aren't that many different ways *in principle* that you could combine a finite vocabulary into an infinite number of different combinations, using sentences that aren't infinitely long. It's true that I could say, 'That cat is very very very very very . . . very big', repeating the word 'very' as many times as I like, but that doesn't count as an infinite number of sentences – at the very least, it's not adding an infinite amount of information to the language. Some clever trick is needed so that we can have non-trivial infinite language from a finite vocabulary.

The first clue to how we make an infinite language is that combining exactly the same words in different ways can mean completely different things. The sentence, 'My dog smells like the babies' has quite a different meaning from, 'My babies like the dog smells'. If I can arrange my 170,000 words in any order, I have at my disposal an astronomical number of potential

---

* This was the lower estimate made by the *Oxford English Dictionary* in 1989, excluding all kinds of derivative and obsolete words.

sentences, and so an astronomical number of meanings. For example, restricting ourselves only to sentences just five words long, there are so many possibilities that you would have to recite 300 sentences each microsecond to get through them all in the time since the universe began with the Big Bang.[*] To all intents and purposes, that is an infinite number of concepts. Problem solved?

The catch, of course, is that no one has a brain large enough to remember the meaning of each of these randomly arranged sentences. This is a universal constraint; even on another planet, full of hyper-intelligent alien beings, there cannot be a unique 'utterance' for every unique concept. If doubling the number of concepts you can express means you need a brain double the size, then expressing an infinite number of concepts would mean an infinitely large brain. Not even mythical sci-fi super-beings, nor even alien supercomputers, could do this.

We need a *clever* trick, not a brute-force one. That cleverness comes through grammar. The sentence, 'Pooh and Piglet go hunting and nearly catch a Woozle', can be arranged in well over 3 million different ways.[†] That's a lot of combinations. But only the tiniest fraction of these make as much sense as the original. They are mainly nonsense like: 'hunting and Pooh nearly and catch a Woozle Piglet go', or, 'nearly go catch and Woozle a Piglet and Pooh hunting'. Very clear rules determine which sentences are meaningful, and which are not. So our grammar both constrains our language and gives it the flexibility it needs to *be* a language. In this way, we can produce different meanings

---

[*] The total number of combinations is $170{,}000^5 = 1.4 \times 10^{26}$, whereas the universe has only existed for about 4×10^{17} seconds.

[†] The first word can be any of the ten words in the sentence, the second can be any of the remaining nine words, and so on, so that the total number of combinations is $10 \times 9 \times 8 \times 7 \times 6 \times 5 \times 4 \times 3 \times 2 = 3{,}628{,}800$.

without our heads exploding from the sheer volume of sentences they must contain.

Chomsky's innovation was to provide a mathematical description of the nature of the kind of grammar that would be able to turn a bag full of words into a rich and infinite language, without being infinitely difficult to remember. What is more, he showed that *only* a very particular kind of grammar would be able to make a language by these criteria. In particular, he felt that the key to making infinity out of finite vocabularies lay in the idea that the *words that come next in a sentence can't depend only on the immediately preceding words*. If words *did* depend only on the preceding words, it would be very restrictive. We would be tied in to using only 'predetermined' sentences. For example, if the words 'Pooh and' were always followed by the word 'Piglet', then Pooh could never go hunting with Eeyore.

The way around this problem is that language is *hierarchical*. I could say, 'Pooh and Piglet go hunting', or I could say, 'Pooh and his pink friend go hunting', or even, 'Pooh and his pink friend with whom he spends much of his time go hunting'. This kind of nested grammar seems to be essential if we are to build an essentially infinite language, without being tied to particular orders of words.

Chomsky organized formal grammars into a hierarchy of increasing complexity, each level being able to generate more complex sentences than the level before.* In doing so, he laid the foundations for what many believed to be a truly universal definition. It was particularly unfortunate though (for our quest in this book) that Chomsky used the term 'universal grammar', to mean an innate grammatical ability present in all humans at

---

* The special kind of grammar that is the minimum required for language is called Context Free Grammar, rather unhelpfully (which is why I don't refer to it by name in the text).

birth. Chomsky himself later made it clear that he didn't mean 'universal' to be 'Universal', i.e. across the universe, at all! But the idea has stuck, as well as the term, and many scientists still believe that aliens *must* use the same kind of grammar as we humans do.

Animal communication researchers have taken Chomsky's grammar with a healthy degree of caution. There are many assumptions that lie behind his reasoning, and if those assumptions aren't met in animals, the chances are they won't apply to aliens either. We need to think very carefully about whether all language, not just human language, must, by definition, meet those same criteria. The idea that language must be hierarchical is very compelling, but for all we know tomorrow's linguists might discover a new way of arranging grammar to be infinitely expressive. When we discover talking aliens, we may find that they have a way of bypassing this requirement for sticking to Chomsky's formal grammars. We should start with the very first assumption that Chomsky used to construct his reasoning: language must be infinite.

Must a language be infinite?

Well, I just said that it does, didn't I? How could we have an infinite number of books if the language isn't infinite? Although this claim seems to make sense, we need to examine this assumption from an evolutionary perspective very, very carefully. Natural selection, of course, doesn't operate with any kind of insight; traits (including language) evolve to become better and better without any particular goal in mind. As no one actually *has* written an infinite number of books, it doesn't seem likely that evolution would favour that infinite ability – it never has the opportunity to be tested! Certainly retrospectively, a language needs to be very large to be a useful language that we or alien engineers can use to write the instruction manual for a spaceship. But how large is very large? And how would natural selection 'know' that a language needs to be big enough for a spaceship manual, when our language first evolved for groups of hunter-gatherers at the beginning of the Stone Age?

There may be a tentative answer to this problem. The key property of a language, even for Stone Age hunter-gatherers, is that it needs to be very extensible. And the best way to do that is probably to make it infinite from the start. As we saw in the previous chapter, birdsong has a huge potential to contain information, but birds don't use that potential. Instead, the huge information-holding ability comes as a bonus with what turns out to be the simplest way to organize birdsong – into notes and motifs. Similarly, a very flexible language is probably easiest to evolve if it's got infinite extensibility in it right from the beginning, even if that infinite ability is never used. If what you really need is to be able to say, 'circle the mammoth on the left', or 'Broud will circle the mammoth on the right', you don't need an infinite language. But being able to make those two sentences, *and to distinguish between them*, may well be done most easily with a grammar that, coincidentally, lends infinite flexibility to the nascent language. We think of the infinite nature of our language as a strict rule, but it might just be an emergent property of a complex grammar.

It is far from clear that every single language, everywhere in the universe, must be infinite. What is clear, however, is that alien languages must have evolved, step by step, from simpler communication, based on the ecological pressures acting on simpler alien creatures on each particular alien planet, and so the communication must have *expanded* to become more complex and more language-like. So extensibility seems to be more fundamental to starting off on the evolutionary road to language, rather than setting our sights on the property of being infinite.

Animals, we keep insisting, do not have a true language. At the very most, we see that they can combine just a few concepts together to make new meanings, but this isn't infinite, or even 'large but not infinite'. If we talk only about the way that animals behave in the wild (you can teach them to do all kinds of tricks in the lab), their repertoire of concepts seems to be limited to just a handful. One of the most sophisticated confirmed cases of animals combining 'words' to give them new meaning is that of the putty-nosed monkey in West Africa. These monkeys have a special alarm cry that goes 'pyow', which they use when they spot a leopard. They also have a different alarm call, 'hack', which is in response to an eagle. But when the monkeys say 'pyow, hack', that means something *completely different*: 'It's time for the group to move on'. Amazing though this behaviour is (it is very rare to identify such complex communication in the wild), it's still a very far cry from an 'infinite' language. It's just three meanings, after all! So we don't see anything among the animals on Earth that is somehow 'in between' infinite, like our language, and fairly trivial, like the monkeys' calls.

Although it's not clear at all that a language really *must* be infinite, it is likely that it will be. Chomsky's constraint of the grammar that enables such infinity certainly sounds like a good indicator of language – if we find infinite grammar in a signal from outer space, it's probably an alien language. But such a grammar is quite possibly not strictly necessary. If we receive a signal that clearly doesn't have the capacity to be infinitely flexible, that doesn't mean that it *isn't* a language. We should keep an open mind that aliens may talk in a way that is very, very flexible, but not quite infinite.

There is one more tantalizing question that crops up when we think about the infinite nature, or otherwise, of language. Chomsky's classification of a hierarchy of grammars doesn't just apply to natural (human, animal, alien) language, but also to computer languages. Although we can speculate that aliens might have a fully functioning but non-infinite language, it's almost certain that a useful *computer* programming language *must* be infinite, and therefore fall into the higher categories of Chomsky's hierarchy. So, if an alien civilization has a non-infinite natural language, would they be able to understand and write software at all? Perhaps they will lack the insight to be able to design and program computers, without which they may never be able to contact us? Fortunately for our search for alien life, these simple-talking aliens are probably not completely excluded from the internet age. Our own human language (*all* human languages) falls into a middling category in the grammar hierarchy, but we are still capable of understanding the concept that more complex grammars also exist. It may take our aliens longer for an alien Chomsky to come along and discover what they are missing, but they would get their internet up and running in the end.

Must language have a grammar at all?

If Chomsky's special grammar that allows words to be combined in infinite variety is not actually necessary for language, surely *some* sort of grammar is required? If we only have 170,000 words in English, we're going to need to combine them in some way, and that way must be orderly. The term 'syntax' expresses a more general concept of how symbols can be combined, and how that combination affects their meaning. We often use syntax to describe, for instance, how notes might be combined in a birdsong, or in humpback whale song. Humpback whales have tremendously complex vocalizations, which nonetheless correspond to a set of rules that dictate how the different components can be arranged. Syntax (in its broadest sense) gives a set of rules that say, for instance, whether two particular sounds can come one after another or not. For instance, an English speaker would instantly recognize 'myanggh' as a nonsense word, whereas Mongolian speakers think nothing of stringing three consonants at the end of a word ('myanggh' means 'thousand').

In fact, in animal communication, syntax is usually fuzzy, rather than strict as in human grammar. Hyraxes sing long songs made up of five different note types, but while the order of the notes is far from random, it's not fixed either. A 'wail' note is twice as likely to be followed by a 'snort' as by a 'chuck', but is rarely followed by a 'tweet'; whereas a 'squeak' is more likely to be followed by another 'squeak' than anything else. It's like the animals are rolling dice in their heads to decide what note to sing next, but the dice are loaded, and it is this loading that makes their syntax. In many ways, the animal world itself is fuzzier than the human world. We understand precise instructions: 'Pick up the green ball and place it in the third bowl from the right.' You might be able to train your dog to pick up a ball and put it in a bowl, but their comprehension of the precise nuances of your instructions will be lacking. Similarly, alien animals evolved from simpler animals with simpler communication needs, so it seems likely that other planets will have ecosystems made up of simpler animals with fuzzier syntax, and more complex animals with more complex communication, and hence more precise syntax.

Syntax distinguishes signals with meaning from random nonsense. It is the syntax of English that shows us that 'nearly go catch and Woozle a Piglet and Pooh hunting' is not a proper sentence. Many people have speculated that looking for signs of syntax, as opposed to randomness, is what will allow us to identify an alien message broadcast to us from outer space. The syntax we see might be in the form of Chomsky's grammar, which would be a pretty good sign that the message was in a proper language, or it may be in the form of a hyrax song, which would imply that there was structure in the message, but would be inconclusive about whether there's language or not. Distinguishing syntax from randomness isn't as easy as it might sound, but it's possible. The statistical properties of a signal are very likely to be different if that signal has been generated according to some (unknown) rules, compared to being generated randomly, but it's not clear exactly how the statistics would be different. Scientists are at this moment working on better and better algorithms to do this, in anticipation of receiving a message that may or may not be from aliens. The Breakthrough Listen initiative* is spending US$100 million over ten years to listen for alien signals, and part of that work involves looking for statistical anomalies which might indicate that a signal from space really does represent a *language*.

Syntax is almost inevitably a part of language. But could there be a kind of syntax that is radically different from everything that we've described and might therefore be immune to the kinds of analyses that scientists are investigating? Considering that aliens may well have some very different ways of processing information from us, we also need to bear in mind that there could be syntaxes which don't necessarily fit the model of human language.

Everything that we've talked about so far treats syntax as a way of ordering symbols – words, in the case of language, or notes, in the case of hyrax- or birdsong. 'Ordering', in this context, means ordering in time; either as words are spoken out loud or as they are read on a page. We start at the very beginning (a very good place to start), then we proceed through the sentence to the end. Temporal linearity is a fact of life for us – just about everything is arranged in time, because time passes inexorably forward. Now, the 2016 film *Arrival* rather fancifully dealt with the question of an alien language in which time is not linear – but unfortunately, as interesting as that may be, it is highly speculative science fiction. So let's consider that on a more solid scientific footing: is there a

---

* https://breakthroughinitiatives.org/initiative/1.

The Last Supper by Leonardo da Vinci.

way in which symbols (not necessarily words) can be arranged according to a syntax, but not arranged in time?

Of course there is, and we see it around us every day. Symbols are arranged in an orderly way in *space*, not in time, and not just as words are arranged on the space of a page (that's really arrangement in time, because you're meant to read them in a particular order). Consider Leonardo da Vinci's painting of the Last Supper. Or the cover of the Beatles' *Sgt. Pepper's Lonely Hearts Club Band* album. These are certainly not just random patches of colour on canvas, and more than that, they are also not just random arrangements of symbols like Jesus and his disciples around the table, or Marlene Dietrich and George Bernard Shaw standing next to George Harrison on *Sgt. Pepper*. Each individual represented is in a certain place on the image, and that place has relevance to the meaning of the work – but these relationships are certainly not relationships in time. Can we say that the painting is a form of language? Is it possible that aliens use painting as language?

Yes it is, and we've met creatures like this before, in Chapter Five – the cephalopods, and cuttlefish, in particular, that communicate using patterns of swirling colours generated by special skin cells. Now, cuttlefish and squid don't actually have a complete language built out of pictures written on their bodies, but neither are their mesmerizing patterns random. In other words, they have some syntax. Although no one has yet analysed these patterns in detail to see how much information could be encoded in a spatial syntax, the mere fact that cephalopods have evolved a communication system on Earth that uses a spatial syntax, means that we absolutely cannot reject the idea that such a language exists elsewhere in the universe. Aliens with a picture language would be difficult to detect, and even harder to understand, precisely because they don't conform to the kind of grammar we're expecting to see in language. But what an amazing possibility – and one that even seems fairly likely!

Language is abstract art

Whether a language is made of words or notes strung together in a time sequence, or images and patterns arranged in relation to each other in space, the power of language comes from combining symbols. We only have 170,000 words, but we combine them in unlimited forms. John Lennon could be standing next to Paul McCartney on the *Sgt. Pepper* cover, but he isn't. The fact that symbols need to be combined to make a language is probably essential. But beyond that, human languages have another important property that may or may not be universal. The words of human language mostly do not have any connection to the object or action that they describe.* The sound of the word 'dog' means a lot to me, but it wouldn't mean anything dog-like to a

---

* Onomatopoeias are the obvious exception to this: words that sound like the object they represent. But these are a very small minority of words in any human language.

speaker of French or Arabic if they didn't already know English, or to an alien, who certainly wouldn't be able to infer the meaning of the word from the sound of it. Words are, mostly, arbitrary representations. Arbitrary, and not just abstracted, like impressionist art. If you squint at a Monet painting, you can see what the painting is about, but no amount of squinting at the word 'dog' makes it look at all like a dog.

For some reason, all human languages have evolved to have this disconnect between words and meaning. What's more, while words have meaning (even if that meaning is arbitrary), the words themselves are made up of really meaningless sounds: phonemes. The 'ee' sound in 'free' and in 'meerkat' have absolutely nothing to do with each other: 'ee' itself has no meaning until it's built into a word. These two phenomena – words being arbitrary and being made up of meaningless parts – are considered by linguists to be fundamental to language itself.

Now, are these phenomena truly *universal*? Can we expect there to be the same disconnect between symbols and meaning across the universe, or is it just a happy coincidence that arose here on Earth? Could an alien language consist of words that all (or mostly) *do* have a connection to the object they describe? Or would such a system not be flexible and extensible enough to be called real language?

Some linguists have attempted to draw up models of the evolutionary processes that could have led to language developing the structure that we have today, but by and large most of the effort in the science of linguistics focusses on languages as we have them, rather than languages as they *might be*. So we're lacking some really strong theories about why words are arbitrary. However, at least two possibilities recommend themselves to us; one specific to Earth and one that is more likely to be universal.

Some linguists believe that the origin of human language was with gestures, not with speech at all.* This argument has quite a lot of merit – if for no other reason than that our closest relatives, the great apes, are very good at sign language and absolutely awful at talking. Although it is not universally accepted that gestures preceded speech, it's a hypothesis with which scientists feel at least generally comfortable. Now, if language really began with gestures, it's no wonder that words are arbitrary, because gestures are largely arbitrary. You can point to your mouth if you're hungry, but there's no gesture for 'dog' that even slightly resembles a dog while being in any way practical. You could get down on your hands and knees and do an impression of a dog, but that is very inefficient. It is possible that all of our assumptions about the essential and fundamental gulf between words and meanings is just a coincidence of the fact that apes have long arms, and not very well-developed vocal abilities. If so, assuming a fundamental gulf between words and meanings is not a basis on which to build predictions about alien language. On other planets, the most intelligent, social and communicative species may not be as long-armed as our own ancestors were.

The second explanation also starts with an Earth-bound observation but ends up being more generalizable. Consider the alarm calls of the putty-nosed monkey. Although 'pyow' is the alarm call for a leopard, it doesn't sound like a leopard. Why wouldn't monkeys make a leopard sound when they see a leopard? That might seem to make more sense – it's certainly an unequivocal signal. However, of course monkeys aren't built like leopards, and can't roar at all, for mechanical and physical reasons. Now, an alarm signal to warn of a leopard should be clear, distinct – possibly frightening, as we saw in the previous chapter – and

---

* The field is inevitably rather technical, but a reasonably accessible textbook and guide is *The Evolution of Language*, by W. Tecumseh Fitch.

unequivocally associated with a leopard. But as monkeys can't actually roar, they must concentrate instead on making the alarm call clear and distinct, and to do that places severe acoustic constraints on the kind of call, so that it will be *different* from all other calls. Imagine that 'pyow' means leopard, and 'pyew' means eagle. That would be fairly disastrous for the monkeys, as half of them might mis-hear the call and run down on to the ground, thinking that an eagle is chasing them, right into the paws of the leopard. The need for *distinct* calls limits what kind of sounds can be used, so that the various calls are very different from each other, and this forces them to be more arbitrary. Arbitrary words may actually be quite common in the universe.

One final consideration about arbitrariness. You might be reminded that the ancient Egyptians had hieroglyphic writing. Although many hieroglyph symbols represented sounds, some of the symbols actually did represent the meaning of the word.★ Is it not possible that aliens could evolve the equivalent of hieroglyphic speech? On Earth, writing evolved long, long after spoken language – but could this be reversed on an alien planet? What about if the language was written onto the bodies of cuttlefish-like aliens that communicate using pictures on their skin? Could this evolve to be a hieroglyphic language? It's a fascinating idea – but is unlikely. In a similar way to the putty-nosed monkeys that need to give a clear and distinct warning about a leopard, cuttlefish hieroglyphics would begin their evolutionary path as very simple symbols, easy for the receiver to interpret. If one alien cuttlefish wanted to warn another about an approaching alien shark, it makes a lot more sense to flash red and black than to draw a beautiful representation of a shark on your body.

★ For example, the hieroglyph ⊓ represents the sound 'pr', but with a vertical line underneath ⊓ the meaning becomes the literal meaning of the picture: 'House' (the word for 'house' in ancient Egyptian is 'per').

The communication would almost certainly start off abstract, and by the time a rich and complex language had evolved, it might sound crazy to the cuttlefish aliens for someone to suggest switching to a hieroglyphic display language. It would be equivalent to someone suggesting that we humans switch to an onomatopoeic language, saying 'woof' instead of 'dog'.

How has language evolved on Earth?

Incredibly, there are few topics in science that are as contentious as this one. Different theories of the evolution of language have their own followers and opponents, and views are strongly held and robustly defended. In some ways, this is not surprising, because language is our one claim to uniqueness as human beings among all the animals, and the precise nature of language, and how it came about, serves to define who we are in a most fundamental way. Like the alien civilizations we hope to meet, we are social and communicative. We have intelligence – even many different kinds of intelligence – but so do many other species on our planet and, we conclude, many other species on other planets (see Chapter Six). In the end, it is our language that makes us different from other Earthlings, and what will make us similar to any alien species that we recognize as being 'like' us.

Although in this book we are not concerned with the specifics of how one particular species on one particular planet evolved language, it is still immensely important to understand the process by which that particular case of language came to exist. It is, after all, the only one we have to study. So although we are less concerned with the specific milestones along the path to human language, we are going to have to delve into the fundamental process that pushed our ancestors along that path.

Similar evolutionary processes may also be operating on other planets.

One of the most striking characteristics of human language, and one that needs to be addressed right at the beginning, is that we seem to be the only species with language on the planet. Of course, we have multiple different human languages, but they are really all variants on a theme – the underlying expressive ability is essentially the same, compared to the difference between human language and that of a hypothetical talking animal, or alien. This is a most extraordinary observation. If language is so great – and look at the accomplishments it has enabled humans to achieve – then why don't other animals have it, no matter how intelligent they may be? We recognize the familiar playful intelligence of the dolphin, the unfamiliar hive minds of the bees and ants, and the unintelligible minds of electric fishes and octopuses. But none of them have a language. How could this be? We can break down our confusion about this unusual state of affairs into three related questions. Is it even true that language has evolved only once on Earth? Is language evolving right now in some species? And could more than one linguistic species coexist together on one planet?

Talking dinosaurs?

Sometimes, no amount of research and empirical observation can separate real scientific hypotheses from speculative science fiction. Perhaps an extinct (and non-human) civilization on Earth did in fact possess a language in the distant past? Would we even know? Treating such ideas as testable scientific hypotheses isn't easy, but our claim of human uniqueness relies on the fact that no other civilization has *ever* existed, and we need some healthy scepticism towards that claim. Might there have been talking dinosaurs? Recently, scientists have tried to put this question on a rigorous footing by examining the current impact of humans on the planet, and asking how a civilization millions of years in the future might detect our long-dead legacy.★ The obvious thought is that our plastics, our concrete and above all our rampant climate change would leave clear fossilized traces in the rocks, easily detected by future geologists. But this is far from certain. After all, it simply hasn't occurred to human geologists *today* to perform the chemical tests necessary to determine whether rocks from hundreds of millions of years ago contain indications of industrial activity.

Perhaps more importantly, though, we see that our own civilization is on a path to planetary alteration on a catastrophic scale. If nothing is done, our damage to the climate and to the diversity of plants and animals will be clearly evident in our geological legacy. Future scientists will look at the mass extinction from this era and compare it to the other great mass extinctions of the past, such as the asteroid impact that killed off three quarters of the species on Earth 66 million years ago. An extinction like that is unmissable in the fossil record. However, it doesn't seem that something like this happened to our hypothetical talking dinosaurs – there's no indication that any of the mass extinctions we observe in the past have been technological in nature. So, either that earlier civilization did not exist, or they did not destroy their planet. And if they, as we ourselves really must do, managed to control their environmental damage, and persist at least for a while in balance with nature, then any geological trace of that civilization would be imperceptible. Perhaps there were indeed creatures with a language before us, but we have not found any indication of them.

★ 'The Silurian Hypothesis: Would it be possible to detect an industrial civilization in the geological record?' by Gavin A. Schmidt and Adam Frank.

Evolving dolphins

The iconic 1960s film *Planet of the Apes* portrayed a world with a fully fledged civilization of talking chimpanzees, orangutans and gorillas, far in Earth's future after human civilization has destroyed itself. Is this likely? Are the intelligent but not-quite-linguistic species of today, like great apes and dolphins, simply waiting in the wings to step into our shoes when we are gone? Perhaps these species are already trudging slowly along an evolutionary road that will lead them to language, and if we wait patiently (several millions of years) then we will share the planet with other creatures that talk and write poetry, and possibly build spaceships too?

There are good evolutionary reasons to doubt this kind of narrative. For one thing, evolution does not push species 'upwards' towards some kind of goal – in this case, language. Each of the intelligent species we think about, chimpanzees, dolphins, etc., are already hugely successful in what they do. They have already evolved to be well suited to their environment, and there is no evolutionary justification for seeing humans, and language, as being 'higher up' than they are. Put differently, there need not necessarily be an evolutionary pressure acting on dolphins, driving them to evolve language. *Planet of the Apes* was a formative film (for me, at least, when I was growing up), but it paints a distorted view of how evolution works. There would have to be a clear fitness advantage to moving along a path to language if any species alive today were to be walking that path. And dolphins and apes don't currently seem to be under that kind of pressure. Something happened, something special, to nudge our ancestors into the path of evolving language. We don't know what that trigger was, but it seems to have been something rare, because as far as we can tell, it doesn't seem to be something that most species experience.

But what *would* happen if dolphins evolved a language? Would they live in peace alongside us, or would there be existential conflict between the two species, with one enslaving the other, as in *Planet of the Apes*? When we discover an alien civilization, will it be made up of a single species, like our own civilization, or might there be multiple different creatures, all playing their own roles in society? Basically, can two linguistic species coexist?

This is a question as much for sociologists as for evolutionary biologists, but the omens don't look good. When modern humans, *Homo sapiens*, walked out of Africa maybe 100,000 years ago, they found populations of other kinds of humans – Neanderthals and Denisovans and possibly others – across Europe and Asia. Within a few tens of thousands of years, only *Homo sapiens* remained. We don't know whether or not Neanderthals had a language, or why they died out. But if they *did* have language then, at least for a short time, our planet hosted two linguistic species at the same time. Only one survived. I'm making an unabashed correlation based on a single sample – not at all the right way to do a scientific analysis – but it is a useful starting point for thought experiments. How would two very different species (that could nonetheless talk) get on with each other?

Evolutionary theory can help us here. It has long been accepted that two species occupying exactly the same niche cannot coexist indefinitely – one must eventually outcompete the other and drive it to extinction. What happens more realistically, and less dramatically – given that evolution operates slowly and gradually – is that species *partition* their niche, so that they can coexist.★ They use different resources so that they are not in direct competition – competition that would always end up favouring one species over the other. For example, the Negev Desert in

---

★ This is known as the Competitive Exclusion Principle and the concept of Niche Partitioning. E. O. Wilson's book *The Diversity of Life* is a good place to start for an overview of such ecological ideas.

southern Israel is home to two related species of rodents, the golden spiny mouse and the common spiny mouse. Both live in the same habitat – in fact the same territories – and eat the same food. Such a situation cannot persist indefinitely; they must learn to share. As it turns out, one species is adapted to forage during the day, and the other to forage at night. Only that way can both species live on top of each other in such a manner.

If two talking species coexist, they are likely to be in competition for the same niche. Even if, say, one species lived on land and the other in the ocean, both would be on a kind of collision course. Language takes us away from the constraints and hardship of the natural world and allows us to hunt cooperatively, build shelters, farm animals and crops, and eventually leave our planet. Linguistic species on the same planet will both evolve towards doing these same things, no matter how different their lifestyles were where they started out. The only way these competing species could survive is if they partition the technological niche. In *Planet of the Apes*, the gorillas were the military, chimpanzees the scientists, and orangutans the politicians. Fanciful as this representation may be, it has its basis in evolutionary theory!

The route to language

So why did language evolve in humans? Well, language is good for cooperating to solve problems. There seems no doubt about this, and it's the first thing that comes to mind when asking, 'Why should language evolve?' Animals that can communicate their problems to others can invite help and build knowledge enabling them to solve more difficult problems in the future. If I don't know how to crack open a nut, I can explain my problem to my friend, who may know that hitting it with a rock is a good idea. If we want to hunt a mammoth, I can explain to my colleagues how to encircle it and what role everyone should have in the hunt. Explanation is so incredibly useful, it seems obvious that it is an ability that should evolve like wildfire in many species, on any planet.

But every trick that improves your fitness comes with a cost. There is always an evolutionary trade-off. The ability to explain my personal survival problems to members of my tribe requires a level of mental processing ability well beyond the capacities of most species. The brain is a very important organ for most animals, and by and large accounts for between 2 and 10 per cent of the entire energy needs of the body. Everyone needs brains, mostly for sensing the environment and giving motion and other commands to the body. Humans, however, have quite outrageously large and active brains; this one organ uses up to 20 per cent of our metabolism. Think about that: a pig uses 2 per cent of its metabolism for its brain, so ten times more of what we eat goes to our brains than does so in a pig. Now, food is a very important and limiting factor for most animals, so there has to be a really good reason for evolution to favour diverting so much of our energy to this hungry organ. Aliens may not have brains that in any way look like ours or have evolved like ours, but they must have a way of processing information, and processing information costs energy. To all intents and purposes, we can call it a brain if it distils the rush of sensory noise that a creature experiences, into useful, concrete information. And just like distilling alcohol from grape juice uses energy, distilling information from noise uses energy. Usually a lot of energy.

Although the idea that language evolved essentially to help humans cooperate to solve problems (hunting mammoth or cracking nuts) makes sense, it doesn't answer the question of how human brains got powerful enough to understand language. There is a chicken and egg situation here. We need powerful brains to talk to each other, but perhaps our big brains are only worth having if we can talk in the first place?

The get-out to our chicken and egg problem is that big brains may have evolved for reasons other than language, and then might have been adopted as useful tools to help us communicate in more detail thereafter. In particular, many scientists think that our large brains are basically social relationship computers. In Chapter Seven we talked about reciprocity, and how difficult it is to remember who it was that actually helped you in the past, so that you can decide whether or not to help them in the future. This is just one of the problems of living in large groups. If you live in a society with a complex dominance hierarchy, it is hard enough to remember who is senior to whom. In chimpanzee groups, it is also important for individuals to form alliances, and manipulate the alliances of others to their advantage. This means that they also need to remember how third parties relate to each other, and how many fights each alliance has won against the other. Maintaining and manipulating such complex information isn't in the behavioural repertoire of most species – it is a special ability evolved only where it gives a particular advantage.

So we can see a plausible scenario where animals living in complex groups evolved large brains to navigate their social environment, and those large brains gave them the potential to communicate ever more complex concepts to each other. Plausible, but not inevitable. Although complex societies seem to be a necessary prerequisite for language evolution, they're not sufficient. Ants and bees communicate very well in their complex societies, without having evolved an infinitely expressive language. But is this social complexity route to language the one that must have been followed on other planets? Are we being too blinded by our own anthropological legacy, so that we can't imagine other ways that language could evolve on other planets?

I think this is unlikely. Language seems inevitably to be a social activity – otherwise with whom would you be communicating? The inherent cost to evolving any complex trait, language included, means that it must provide some *immediate* benefit. Evolution doesn't work on 'might-be's, potential future benefits from having a language that might come several generations down the line. It is not an Earth-centric assumption to say that language will only evolve in social animals. But sociality itself is a complex trait, and similar trade-offs to the ones between social savvy and brain size described above are likely to be universal. Although as yet unimagined alternative routes to language may be possible, ours seems to be objectively reasonable. There's a good chance that at least some alien civilizations will share an evolutionary history of language like ours.

A language signature

Astronomer Carl Sagan's epic science fiction novel *Contact* has the protagonist, Dr Ellie Arroway, receive a radio signal from outer space that appears to contain a sequence of prime numbers. But why? Why would anyone send us a long list of numbers? She explains:

> This is a beacon. It's an announcement signal. It's designed to attract our attention . . . It's hard to imagine some radiating plasma or exploding galaxy sending out a regular set of mathematical signals like this. The prime numbers are to attract our attention.

Ellie was lucky (if fictional): she got a beacon. Sagan had long wrestled not just with how to listen for alien messages, but also with how to send them. Prime numbers, he thought, were the key, as there is no known natural process that can generate a sequence of prime numbers. It's an incontrovertible sign of an intentional signal. So when the day comes that we receive such an alien signal, we would undoubtedly be able to recognize a beacon.

But what if we receive a signal that's not intended for extraterrestrial ears (i.e. us)? The radio and television broadcasts of planet Earth are expanding into space at the speed of light, as an everinflating bubble around us. The potentially habitable exoplanet TRAPPIST-1e, just thirty-nine light years away, is, as I write these words, receiving our news broadcasts about the failed American attempt to rescue hostages from the US embassy in Tehran in 1980. What would any inhabitants of the TRAPPIST-1 system make of our signals? These signals are certainly not a list of prime numbers – would the Trappists even recognize that this is intentional communication? Would *we* recognize alien television broadcasts?

In fact, it would likely be pretty easy to recognize technological broadcasts, even if they're not prime numbers. There are plenty of fingerprints of technology, in terms of patterns that don't occur in nature. Even if alien technology is very advanced, we are confident we will be able to recognize it as technology. But what about the *language* that a signal contains? Do we have a way of recognizing language? If we were to land on an alien planet and hear a green alien creature babbling to us, would we be able to determine: this animal is speaking in a language, as opposed to the alien equivalent of birdsong? Can we even know that animals on Earth aren't speaking to us in a language?

What we are looking for is a universal fingerprint of language – some kind of mathematical algorithm that processes a signal and says 'yes, this is' or 'no, this isn't' a language. We need to do this before we can even begin to think about translating any signal. But is it reasonable to think that such a thing as a universal fingerprint might exist? We have already explored some pretty outlandish ways that aliens might communicate, like swirls of colour on their cuttlefish-like skin. How could any algorithm that we develop based on *our* language identify *theirs*?

The most common approaches to finding a language finger-print candidate are indeed based on our own language. If a language is made of words and sentences, then there are several ideas based on a branch of mathematics called information the-ory, which give us various predictions about the statistical properties of those words and sentences. The underlying idea is that language should be complex enough to express all the con-cepts that you need, but not so complex that unreasonably large brains (or their equivalent) are required to produce and interpret the sentences. There is a trade-off between complexity and sim-plicity, and any set of sentences that are 'well balanced' in this trade-off are a good candidate for linguistic sentences.

One way to measure where a signal is on the spectrum between complexity and simplicity is to see how common the most common words are. It is a well-known but peculiar fact that the most common word in English ('the') is twice as com-mon as the second most common word ('of'), and three times as common as the third most common ('and'). In fact, this relation-ship is pretty well preserved for the top 10,000 words in English, with word number 10,000 being one ten thousandth as common as 'the'. Even more curiously, this effect is observed in every other human language as well! In French 'le' is twice as common as 'de', and three times as common as 'et'. Make no mistake: this is a very peculiar observation.

Now, this phenomenon, known as Zipf's Law after the Ameri-can linguist George Zipf, who formalized the observation in the 1930s, has gained a lot of attention among scientists working on the Search for Extra Terrestrial Intelligence (SETI).* If Zipf's Law is

---

* I should mention that recently there has been an explosion in new compu-tational methods for detecting language signatures – Zipf's Law is only one approach. Some more are discussed in *Xenolinguistics: Toward a Science of Extraterrestrial Language*, edited by Douglas Vakoch.

truly a universal property of language, it would be a good rule by which to measure any signals that we receive. Unfortunately, the precise reason why Zipf's Law applies to languages is still slightly unclear, and this means that we can't yet say with certainty whether all languages, not just Earth-bound ones, should follow the same law. But there is some logic behind that claim.

Essentially, Zipf's Law indicates a balance between complexity and simplicity. Consider a truly random signal made up, say, of the first five letters of the alphabet: A, B, C, D and E could come in any order and with equal probability. No matter what sequence of letters I have already received, there's a one in five chance that the next letter will be an A, and an equal probability for every other letter. This kind of signal is not just random, it's very complex. It's as complex as it could possibly be, because you have not the slightest way of knowing what letter will come next. In information theory, complexity and randomness are almost the same thing. That sounds counterintuitive, because we expect an intelligent message to be anything other than random. But information theory and Zipf's Law tell us nothing about what someone *intended* to put into a message, or even about how much information is actually *contained* in the message, only about the *potential information capacity* of that message. A random signal could, in theory, contain the most information although, of course, if it is truly random, it contains none. It's like taking a large text file and compressing it using a file compression algorithm. That great big file becomes tiny, but the characters in the compressed file appear random – because that's the most efficient way to store the information.

If you find that hard to believe, consider the opposite case. In this signal, 96 per cent of the time the letter A arrives, and only 1 per cent for each of B, C, D and E. Whatever happens, I am almost positive that the next letter will be A. How much information is in that signal? Not much – basically just 'A'. These stereotyped signals, where you can guess what's going to come next, are simple but contain little information. So the spectrum between complexity and simplicity is really a spectrum between randomness and repetitiveness.

When you do the maths, it turns out that right in the middle, in between the case of 20 per cent for each of the letters (random), and 96 per cent vs 1 per cent (stereotyped), lies Zipf's Law. That balanced case would predict that A was twice as common as B, and three times as common as C, and so on. For a signal consisting of just five letters, that works out approximately as 44 per cent A, 22 per cent B, 14 per cent C, 11 per cent D and 9 per cent E. It would seem that these probabilities represent the *objectively* balanced way to send a signal: not too complex, not too simple. That's why – so the story goes – Zipf's Law is so widely observed. It provides just enough complexity for information, but not so much as to be overwhelming.

Now an interesting thing happens when we test animal signals for Zipf's Law. They almost all fall on the 'simple' side – too stereotyped to be language. And this makes sense: birdsong does sound rather more repetitive than Shakespeare. Beautiful as it is, birdsong could never be language – it is too simple, too repetitive, it doesn't correspond to Zipf's Law, and it doesn't hit that balance between complexity and simplicity. However, some animals fare better. A colleague of mine, Laurance Doyle of the SETI Institute, one of the first to propose using animal communication as a testing ground for recognizing alien signals, found that dolphins have communication that corresponds as closely as one might expect to Zipf's Law. In my own research, I've found that orca also have this property. Intriguing.

However, this definitely doesn't mean that either dolphins or orca have language, and this is the biggest problem with looking for fingerprints of language, and with Zipf's Law in particular. There *must* be a lot of false alarms. For one thing, the balance between simplicity and complexity that Zipf's Law identifies isn't just necessary for language, it seems to be a generally good thing all round. Maybe other species find this balance suits their needs that are not linguistic. Orca certainly communicate complex information, but it doesn't have to be language to benefit from Zipf's Law. In fact, some other species, including some songbirds of whom you wouldn't expect great linguistic feats, also follow Zipf's Law to a certain extent. I would speculate that Zipf's Law may be a precondition for a communication system before it can evolve to be a true language, but on its own we can't use it to test for alien language. There are many other reasons why it might apply.

The real limitation of Zipf's Law is that it only measures word frequency, and language is about a lot more than just how often the word 'and' appears. Information is contained in the relationship between words, as well as within the words themselves. As we saw, 'Pooh and Piglet go hunting and nearly catch a Woozle' is a true linguistic sentence, whereas 'nearly go catch and Woozle a Piglet and Pooh hunting' is not, although they both would correspond to Zipf's Law. It is possible that we will discover an equivalent of Zipf's Law that applies to the complexity of a grammar, but we haven't found anything yet. Unfortunately, from my research it appears that even human language grammar is more complex than Zipf would expect, and so our search for a universal language fingerprint continues. Nonetheless, at least Zipf's Law gives us a first filter; a way of ruling out a signal if it is too simple or too complex to be language.

All of our language fingerprint research has so far focussed on sequences of words or symbols. This is, of course, how our language works and, far from being an immodest bias, it's simply that we don't have any other kind of true language against which to test our algorithms. It's no good designing a test for the fingerprint of a cuttlefish image language, because we can't say what a true image language would look like. However, this is a field of active research, and it's reasonable to assume that some form of information theory will apply to image languages, just as it does to sequence languages. Images can, after all, be compressed into random sequences, just like text files, and that means that their complexity can be measured and quantified just like the complexity of sentences. These ideas are, to me, among the most exciting avenues for SETI research, without ever seeing an actual alien.

We began this chapter saying that we don't know what language is, and we haven't come to an answer. But that's OK. Our attempts to understand what alien life is like is as much a quest to find the right questions as it is to find the right answers. Our future neighbours and hopefully our conversation partners will have a language of some sort, and it would be arrogant to claim that we can know in advance what kind of language that will be. But it would also be irresponsible to wash our hands of the question and say, 'we know nothing'.

In fact, we can be confident of some of the fundamental properties of alien language, no matter how different from ours it is. Language has two critical features that are shared everywhere: it is a way to communicate complex concepts, and it evolved by natural selection. When we look at all the possible ways of achieving language, it is clear that no one communication system has a monopoly on meeting those criteria. Still, many communication systems fail one or the other of those two tests. Either they do not have the expressive capacity to be classed as a 'language', or they do not lie on a realistic evolutionary pathway – that is, we cannot see how that language could evolve step by step, with each step providing a distinct small advantage. When trying to identify, and eventually translate, alien languages, we will be better served by looking for these fundamental features, rather than searching for direct parallels between our own language and theirs. Unlike translating an unknown human language, which we can be confident is made up of words like nouns and verbs, with an alien language we will need to ask different questions before we can begin to translate. Where is the information contained? What purpose does the language serve? How did this alien language co-evolve with alien social intelligence?

Interestingly, the conclusion is that many of the things we assumed were fundamental to language (words and sentences, grammar) are not so essential, whereas things that we never thought of as being particularly 'linguistic' (social living, say) are likely to be as essential on alien worlds as they are here. Language is such an integral part of being human that we assume that other conscious and intelligent beings share this spark of humanity. Indeed, language does, in a profound way, define both humanity and the alien equivalent of humanity. But that fundamental essence of language shared by all civilizations across the universe is something much deeper than the ability to say, 'Live long and prosper'. Although I still hope we'll be able to make a good translation of that when we make First Contact.

10. Artificial Intelligence – A Universe Full of Bots?

Throughout this book I've been stridently insisting that natural selection is the only way that complex life can evolve. Now it's time to confess: that's not strictly true. There is one other way. Life could be designed by some other, intelligent, life. I'm not talking about divine creation, but rather about that hazy near-future when the human race will be able to make artificially intelligent robots and machines that could in many ways resemble natural animals and plants. We can already design computers that can learn and reason and – almost – convince an observer that their behaviour might be human. It's not unreasonable that in 100 or 200 years, our computer systems will be effectively sentient: human-like robots, similar to *Star Trek*'s Commander Data. Alien civilizations that are considerably more advanced than us are likely already capable of such creations.

The possibility – likelihood, even – of such robotic life has implications for our predictions about life on alien planets. If, as some astrobiologists believe, alien life is likely to be artificial – i.e. 'manufactured' – would the rules and constraints that we have discussed in the last nine chapters still apply?* Or perhaps there are different rules, and different constraints when life is the product of clear, intentional design?

Natural selection seems, at first glance, to be so frustratingly inefficient. Generation after generation of baby gazelles are born, destined to be eaten by lions. Only by chance is one baby

---

* See the paper 'Cultural evolution, the postbiological universe and SETI', by Steven J. Dick.

born with longer legs, able to run faster, and so escape being eaten. Generations of flies are eaten by birds – then a chance mutation gives one of them yellow stripes that scare the birds away. Why should all of this be left to chance? Wouldn't evolution be so much faster if gazelles could *know* they need to be faster, and flies could *engineer* themselves to be scarier? Of course, the very beauty of natural selection is that it doesn't require any foresight; natural selection explains life in the universe precisely *because* there is no presumption of any prior knowledge. No Creator is necessary, because the evolutionary process is guaranteed to proceed even without any predefined rules. Life evolves – albeit slowly – without having to know where it's going.

But what if it were all different?

What would life look like if it *did* know where it was going?

Crabs on the Island

The 1950s physicist Anatoly Dneprov wrote quirky and characteristically Soviet science fiction. His novel *Crabs on the Island* tells the story of two engineers conducting an experiment in cybernetics on a deserted island. A single self-replicating robot (a 'crab') is released, and forages for the raw materials to build other robots. Soon the island is overrun with baby robot crabs. But the crabs begin to mutate. Some are larger than others, and ruthlessly cannibalize the smaller robots for spare parts to build even larger robots. How would such an experiment end? Catastrophically, of course, as is consistent with the genre.

The 1950s was an era full of ambitious expectations from science and engineering. Although visions of flying cars and personal robots were overly optimistic, many of the ideas floated during that time were reasonable concepts, even if technology was much further behind reality than people expected. In 1956, an article in the popular magazine *Scientific American* talked about the possibility of solving the problems of food production by creating artificial robotic 'plants' that would absorb nutrients, synthesize food chemicals, and then cheerfully surrender themselves for harvesting – in the words of the article, 'Like lemmings, a school of artificial living plants swims into the maw of the harvesting factory.' Crucially, such artificial plants would also have to *reproduce*. Just like naturally living organisms, they would have to use natural materials to build new copies of

Like lemmings, a school of artificial living plants swims into the maw of the harvesting factory.

An illustration accompanying the 1956 article on artificial plants in *Scientific American*.

themselves – copies that would reliably continue to perform the same functions of fattening themselves up for harvest.

In principle, the idea is sound. Natural selection has given the world a huge array of plants and animals, most of which are not particularly tasty to humans. When we understand how life grows and reproduces, why not use this knowledge to build our own creatures – creatures that suit our needs more closely, without the constraints and baggage of their evolutionary history? If we want to eat meat but don't want to cause pain and suffering, surely we can design machines that grow 'muscle' without a brain or nervous system? Even if it sounds uncomfortable that such machines 'swim like lemmings into the maw of a harvesting factory', their behaviour is no more than a software program we have given them. There is no more suffering involved than when you throw your old smartphone into the bin.

Human technology has begun to move along this route, for instance using genetic engineering to insert particular genes into other organisms and create, for example, bacteria or yeast that produce insulin – life forms hijacked for our own purposes. Growing meat in an organism without an animal brain is quite clearly within the range of our realistic imagination. But what might we expect of alien civilizations, possibly thousands of years more advanced than we are? Are their worlds likely to be filled with robotic cows and sheep, producing milk and meat from complex miniaturized chemical machinery? Or perhaps such civilizations will consist solely of artificial organisms, feeding, reproducing, fighting and cooperating, with no 'naturally' evolved creatures in sight. Perhaps carefully designed artificial plants and animals will work more efficiently than clunky naturally selected organisms like us – so much so that 'native' aliens will be outcompeted and replaced by their own artificial creations.

Most sci-fi tales of self-replicating machines are dystopian in the extreme; well-meaning intentions to create artificial life end with exponentially growing numbers of 'bots' swarming over the universe, converting every planet and every star into more and more copies of themselves. Even a simple bacterium could spell the end of the universe, if it continues to grow exponentially without limit. The bacteria *E. coli* can, under ideal conditions, divide into two daughter cells in about twenty minutes. Starting with a single organism, in an hour there would be eight. In a day, there would be 4,000 million million, weighing 4,000 tonnes. In just seventy-two hours, the mass of *E. coli* would be greater than the mass of the entire universe.★ Such is the curse of exponential growth.

Of course, this doesn't happen. Science fiction can be terribly pessimistic, but that pessimism is unfounded. Other factors are at play. Resources are limited. Eventually, even the crabs on the island run out of materials with which to make new robots. On top of that, with so many tasty bacteria around, other creatures evolve to eat them and an equilibrium is reached. We don't need to worry – the universe is not going to end in bacterial slime. Admittedly, humans have caused tremendous damage to our own planet, but we've hardly destroyed the universe. In fact, there's no indication in the night sky that any organism, biological or artificial, has spread its influence as far and wide as we might expect if they were growing exponentially like bacteria or robot crabs.

But we must also be cautious with our optimism. We rely on the age-old processes of natural selection to keep reproducing bacteria or robot crabs in check; something will evolve to eat them. But what if these were not bacteria but intelligent organisms, plotting a way to find new resources, discovering new

★ After n generations of twenty minutes, there will be 2^n bacteria, each weighing 10^{-12}g. The mass of the universe is about 10^{56}g, which corresponds to just 2^{216} bacteria, or 216 generations: seventy-two hours.

ways to improve themselves, their evolutionary fitness and their ability to learn from each other and from previous generations? Could such an army of replicating artificial intelligences be possible? If so, could they be stopped? How realistic is it that alien planets may be inhabited by artificial creatures so advanced that they can bypass natural selection itself? And if that is possible, why has such a creature never evolved naturally? If we want to know whether or not we should fear alien artificial intelligence, first we have to understand what's so special about it.

Jean-Baptiste Lamarck and the giraffe's neck

One of the reasons why evolution is so slow on Earth is that, in general, children are not born with the benefit of their parents' experience. Baby gazelles instinctively know how to run from lions, but that running instinct has evolved painfully slowly over the generations, with those babies that were lacking the instinct simply becoming lion snacks. But imagine how much quicker evolution would be if the first mother gazelle to escape a lion had babies that all feared lions. By and large, that doesn't happen. Recent discoveries have uncovered possible mechanisms for extreme experiences (such as famine or disease) having an effect on future generations, but it certainly doesn't seem to be a major factor in natural selection or evolution.* Why not?

Before scientists understood the nature and mechanism of

---

* Called 'transgenerational epigenetic inheritance', the importance (or otherwise) of this mechanism in actual evolution is still very controversial. There are a number of popular books on the topic, such as: *The Epigenetics Revolution: How Modern Biology is Rewriting Our Understanding of Genetics, Disease and Inheritance* by Nessa Carey; and *Epigenetics: How Environment Shapes Our Genes* by Richard C. Francis.

inheritance, it was considered at least a reasonable suggestion that the experiences in an animal's lifetime would be passed on to its offspring. This idea is generally associated with the Enlightenment French biologist Jean-Baptiste Lamarck (1744–1829), who, in the century before Darwin, attempted to explain the fact that animals appear to be remarkably well adapted to their environment. How could this come about? Jean-Baptiste had a number of ideas how this might happen, but is best known for his two-pronged laws of inheritance. Firstly, that animals develop traits that they use repeatedly, and lose traits they don't use; thus moles became blind, because they didn't use their eyes underground, and, famously, giraffes developed long necks because they were stretching up to reach high leaves. Secondly, and crucially, Lamarck proposed that animals can pass these acquired traits to their offspring; if a mother dog sees a snake and gets scared, her puppies will also be scared of snakes.

What has come to be called 'Lamarckism' we now know to be broadly incorrect, and it has led to a considerable degree of unfair ridicule for Jean-Baptiste. A German biologist, August Weismann, attempted to disprove the law of inherited traits by cutting off the tails of mice, generation after generation, looking for a mouse born without a tail. It is hard for us today to take such experiments seriously, particularly as at least 100 generations of Jews have been circumcised at birth, without any of their offspring being born without a foreskin. Oh well, one of the things about science is that sometimes you just have to do the experiments to convince yourself that the experiment was pointless.

Today, we actually understand the molecular mechanism of inheritance. We now know that, for the most part, neither of Lamarck's rules are correct on planet Earth. But it's not at all clear that they *must* be incorrect. That is, must they be incorrect on every planet, no matter what kind of biochemistry the organisms use to build their bodies and to reproduce? Perhaps incorporating experiences into the alien equivalent of genes isn't as difficult as it is with DNA. It's fair to say that if there *is* a planet where Lamarck's rules apply – or if the inhabitants of another planet were to design artificial life that could pass its experiences to its offspring – then evolution would take a very different, and probably unrecognizable, course. Animals and plants would adapt so much faster if they could pass their experiences directly to their offspring!

What, if anything, is fundamentally wrong with Lamarck's ideas? Biologists cannot say with certainty. We can say that they're wrong here on Earth, but that could be no more than a coincidence of the chemical mechanisms by which DNA reproduces itself into new baby animals. There are two arguments, however, which indicate that Lamarckian evolution is unlikely to be the natural state of affairs on other planets.

Development, development, development

Most people have a general understanding that natural selection works by random mutations. Favourable mutations survive and spread through the population, whereas unfavourable mutations mean the unfortunate owner dies quickly. However, that's not really an accurate explanation of the evolutionary process. Most major mutations – the vast majority – are very bad for you. If we had to wait for favourable mutations to give us eyes, or wings, or long necks, we would be waiting a very long time indeed. Most favourable adaptations actually take place not as a 'mutation for an extra arm', but as a result of much simpler mutations that change the way the embryo develops. Subtle influences are powerful, and less risky. You can make clay pots of hugely different shapes and sizes on a potter's wheel by subtly varying the force you exert on the clay as it spins. This is far more beautiful than just slapping on an extra lump of clay – despite the admiration we show for the artistic creations of our primary school children who do just that.

Consider Lamarck's most famous example: the long neck of the giraffe. If you are a short-necked giraffe ancestor, how can you reach the high leaves that no one else can reach? There is a chance that your baby giraffe would be born with a special mutation – an extra neck bone! Lucky Giraffe Junior would be able to reach those unexploited leaves, she would be more likely to survive and reproduce than her companions and would pass on that same mutation to her babies. And many animals do occasionally have mutations that provide them with an extra vertebral bone, but giraffes have the *same* number of neck bones as a human, or a mouse. A major mutation is not just unlikely, it's also risky. Such a significant skeletal change would have to go along with changes to other parts of the embryo's development: all the nerves and blood supply would have to change simultaneously, too, otherwise young Junior would be unable to function as a giraffe at all. Big mutations aren't usually good for you. Sudden changes are more likely to harm than help.

The other possibility is that a much simpler mutation takes place; a mutation that affects how the baby giraffe grows. Maybe she starts growing a neck earlier than other giraffes' embryos. Or her neck bones grow faster, or for longer. There aren't more neck bones: each one is just bigger. By the time this embryo is born, she will have a slightly longer neck, without having made any drastic changes to the basic body design, and this is indeed how giraffes have long necks. In recent decades biologists have come to an understanding that these subtler processes are more likely to be driving the adaptation of species, rather than sudden, dramatic advantageous mutations.* As in real estate, where the

---

* See *Endless Forms Most Beautiful: The New Science of Evo Devo and the Making of the Animal Kingdom* by Sean B. Carroll.

three most important factors are 'location, location and location', it would seem that in evolution the easiest adaptations to make are in 'development, development and development'.

But this, of course, is the mechanism on Earth, and doesn't necessarily say much about how animals evolve on other planets. Yet there is an element of universality here. It doesn't really matter exactly what the mechanism of inheritance is; it could be DNA, or some equivalent molecule, or even some unimaginably different process to ours. Whatever the process, we could argue that sudden changes in functionality are unlikely to be beneficial. However alien species reproduce, we shouldn't expect their offspring to be radically different from their parents. As they say, 'If it ain't broke, don't fix it.'

One of the most intuitive ways of visualizing this process has been popularized by the theoretical biologist Stuart Kauffman.* Let's use the metaphor of climbing to the top of a mountain in heavy fog as we did in Chapter Two. The higher your altitude, the better adapted you are to your environment in evolutionary terms; the longer your neck if you are a giraffe, or the faster at running if you are a gazelle. How do you get to the top of the mountain, to peak adaptation? You could climb slowly, following what seems to you to be an upward path – that should be a good strategy – but if I gave you a magical teleport device called a 'mutator' that moves you instantly 100 metres in a random direction, would you use it? Well, that depends on the terrain. If you are in the flat Cambridgeshire fens, you probably should. You're unlikely to be anywhere near a hill, and so jumping around is as good a strategy as any. But if you're climbing in the Lake District, or the Smoky Mountains, you're better off crawling in a consistently uphill direction. On a mountain, up usually

* See *At Home in the Universe: The Search for the Laws of Self-Organization and Complexity* by Stuart Kauffman.

takes you up; most of the time, jumping around is more likely to lead you astray than to help you reach the peak. Evolutionary landscapes are (for reasons explained in more detail in Kauffman's book) most likely to be rugged but gradual, like the Lake District. Keep climbing. Don't teleport. Don't mutate. This is a rule that could apply on any planet.

The showdown: Lamarck vs Natural Selection

The second reason to think that Lamarckian evolution is unlikely to be the norm on other planets comes from computer simulations. It's so tempting to think that inheriting experience is useful – how could more information be worse than less information? Nonetheless, we can't make assumptions like these without testing them. So what do simulations show? Scientists have implemented computer versions of evolutionary worlds with lots of virtual software creatures called 'agents' competing against each other for virtual resources.* The agents are given a very limited artificial intelligence, a neural network, and are left to learn about their virtual environment. They then mutate and evolve according to two different rules: one set of virtual creatures evolves by natural selection ('Darwinian' agents), with particularly successful agents becoming more numerous, whereas the other set ('Lamarckian' agents) can pass on their learned neural net to their 'offspring' – in other words, their offspring start off life with the information that their parents' neural net has learned. Which will be more

---

* *Artificial Life: The Quest for a New Creation* by Steven Levy is a general introduction to the topic, but the specific study about Darwinian and Lamarckian agents is T. Sasaki and M. Tokoro, 'Comparison between Lamarckian and Darwinian Evolution on a Model Using Neural Networks and Genetic Algorithms' in *Knowledge and Information Systems* (2000) 2:201.

successful? Put them in a cage-fight: who will win? Lamarck or Darwin?

The answer is, 'it depends'. In a relatively constant environment, being able to pass your experience to your offspring is more beneficial. The Lamarckian baby agents begin their existence already knowing a lot about how to find and exploit resources, whereas those that don't inherit their parents' experience have to learn everything about the world from scratch. Lamarckian evolution – if there is a mechanism for it – is more effective. Being born smart sounds like a winner.

But in a changing environment, the tables are turned. Being born knowing about the world can be a disadvantage if the world is always changing. Lamarckian babies set off convinced that they're doing the right thing, only to find that they've been given bad advice. The leaves on the top of the tree aren't tasty anymore, they're poisonous. Now what are you going to do with your long neck? Like Kauffman's teleporting mutator, they find themselves in unfamiliar terrain with no obvious way to un-learn their now useless knowledge.

Conversely, and somewhat confusingly, in a changing environment, you might expect Lamarckian babies to be at an *advantage*.* Suddenly a new predator appears, and within a single generation, the offspring of parents who had a narrow escape will be more wary. This is an advantage. Traits that can be transferred from parents to offspring without waiting for a rare genetic mutation will spread faster through a population. However, in a rapidly changing environment, jumping on every bandwagon is a risky business. Like investing in every new cryptocurrency that comes along, you are more

---

* Again, this is a hugely technical field, but for a technical review of the problem, see 'Epigenetic Inheritance and Its Role in Evolutionary Biology: Re-Evaluation and New Perspectives' (2016) by Warren Burggren.

likely to be heading down a dead-end avenue than a road to riches.

It's true that these kinds of computer simulations and thought experiments only tell us about computerized and imaginary environments, and cannot really tell us facts about strange alien worlds that actually exist. We can't be sure that Lamarckian inheritance will be out-performed by natural selection in a changing environment on a real planet. But it gives us important food for thought. The evolutionary history of the Earth has been one of massive environmental changes. The asteroid that spelled the end of the dinosaurs was nothing compared to the upheavals that life has endured over the past 3.5 billion years. At one point, the oceans froze over from equator to pole, which clearly posed a huge challenge to all life, and only a few adaptable organisms survived. We have no knowledge of the climatic history of other planets, but if it's anything like it has been on Earth, perhaps those Lamarckian organisms that doggedly persisted with their inherited experience may well have died out early in the planet's history. By the time the biosphere calmed down and became more reliable, perhaps only the plodding, cautious, Darwinian natural selectors survived. Any life forms that thought they knew what they were doing got caught out when the rules changed.

The key thing is that it is not enough simply to inherit experiences. That risks being decidedly disadvantageous in a rapidly changing environment. The organism needs to be able to *know* when to use those inherited experiences and when to reject them. That knowledge, or intelligence that we would expect artificial life forms to possess, is crucial to leveraging the ability of Lamarckian inheritance. But such a mechanism seems implausible in early life forms like bacteria, as that kind of decision-making implies a level of information processing quite beyond the capacities of simple life forms. In fact, it requires a decision-making apparatus: a brain, in essence. Perhaps Lamarckian animals – if they ever existed – couldn't keep up with the changing environment, and died out long before they had evolved the brains they needed to take advantage of their latent superpower.

So far, this has been largely speculation. I tend to think that inheritance on other planets will be similar to that on Earth, but neither would I be surprised to find that Lamarckian mechanisms also existed on some worlds. But surely we, as intelligent creatures, can do better than dumb software agents? Surely we can design something that works well in both constant and changing environments? Or if not us, surely a more intelligent alien race may be designing artificial organisms right now, and sending them out to colonize the universe?

A better solution

So the possibility of Lamarckian organisms arising naturally is at best moderately implausible, based on the arguments above. It's clearly advantageous, but problematic, for my offspring to benefit from my experience, but natural selection can't incorporate the benefit of experience into the next generation. Could such organisms be created artificially? And what might such artificial organisms be like? If we were to design such a creature from scratch then we would want them to learn from experience, because that clearly gives them an evolutionary advantage, but we want them to do it in a more intelligent way than the simple agents in the simulation. Actually, we would want them to do it rather like *we* do, passing *our* knowledge from generation to generation. Humans are actually rather good at keeping knowledge and passing it on, and also at adapting our knowledge for changing conditions, unlike the poor Lamarckian software agents.

Creatures created artificially, with the intentional purpose of accelerating natural selection, should have this elusive super-Lamarckian property. They would be able to convey information from one generation to the next, without getting stuck in dead-end adaptations as soon as the environment changes. They should be able to reason, predict and pass on traits that are helpful, while simultaneously rolling back adaptations that are no longer useful. Artificial organisms would not have an appendix to become infected, no wisdom teeth to be painfully extracted, and certainly not a birth canal too small for babies with large brains.

All mammals have a laryngeal nerve that supplies brain signals to our larynx, controlling our vocal cords and allowing us to roar, squeak and talk. For essentially arbitrary reasons, that nerve loops around one of the major blood vessels near the heart. While this tiny detour was of no consequence for our fish-like ancestors, some animals like the giraffe have evolved longer and longer necks, so that the larynx gets further and further away from the heart. In the giraffe, that same nerve makes a four-metre journey from the brain, all the way down the neck, round the same blood vessel as in frogs and mice, and all the way back up the neck again to the larynx. Any organism capable of refining its own design would dispense immediately with such an anomaly. Our hypothesized creature would be able to adapt both forwards and backwards: forwards in the sense of predicting what adaptations would be beneficial to them in the future, and designing such adaptations into their bodies; backwards in the sense of identifying useless or harmful parts of their bodies and removing them from future generations. An organism that could do this would be well placed to take over their world. An artificial creature designed by super-intelligent aliens would surely have such a capability built in.

The Cultural Revolution

It has probably not escaped your attention that the kind of intelligent transmission of experience from one generation to the next – together with the ability to know when to use that information – is not unlike what we see in human society in the cultural transmission of ideas from generation to generation. We don't need genetics to learn about science, we just need a school. More importantly, we don't need to follow a religion or a political ideology indefinitely and unchallengingly, we can detect when it's not serving our needs and change our direction. Cultural transmission of experiences is a process with spookily Lamarckian characteristics.* We do indeed inherit a tendency to certain cultural ideas from our parents and from society, but we can mould them to our best advantage, alter them, or even discard them. You might be brought up by parents who are wonderful musicians, but you decide that you never even want to touch a kazoo. Cultural ideas that are used are reinforced, those that are neglected waste away.

Imagine if you could choose to say, 'You know what, Mum? I don't think I'll have a cervix too small for my baby's head after all.' Well, you can't, because your physical body evolves by natural selection. But you *can* say that you don't feel like managing without drugs during labour like she did – or conversely that you don't want to take drugs during labour like she did. Cultural transmission of ideas empowers us in almost every aspect of our lives.

Today we are accustomed to the idea of cultural transmission and it seems natural to us, but humans have only had language – and thus the ability to explain ideas – for the last two or

* See *The Meme Machine* by Susan Blackmore for a detailed discussion of the way that cultural memes do (and how they do not) evolve.

three hundred thousand years or so. Is this a peculiar anomaly that we shouldn't expect any alien species to possess? Indeed, are we the only examples in the history of our planet to have this exceptional ability to spread adaptations by communicating them, rather than by genetics? Certainly not. Birds learn songs from each other, and in many cases, cumulative cultural transmission leads to arbitrary 'dialects' in birds from different geographic regions. More advanced even than that are the several species of birds, including crows and tits, which have been shown to 'copy' new ideas (mostly new ways of finding food) from other individuals, eventually leading to something that resembles a 'viral' culture spreading through the population. Famously, in the 1920s blue tits in England began to figure out that they could access the cream inside milk bottles delivered to household doorsteps, by pecking a hole in the card or foil cap. This behaviour spread quickly across the whole country, with those birds not yet in on the secret observing those who had already learned the trick. Dolphins in one location off the coast of Australia teach their daughters (it's mostly just the girls that do it) to cover their snout with a special kind of sponge, so that they can dig around for food in the seabed gravel without hurting their snouts. Cultural transmission exists in the animal world around us, it's just not quite as powerful and as detailed as ours.

You can easily argue that our entire civilization, and all of our technological and artistic achievements, would have been impossible without the ability to pass our ideas to others (not necessarily our children), for them to develop and improve. Science and technology are built on layer after layer of cumulative knowledge, refined and improved as it is passed from individual to individual. When necessary, unhelpful (or plain wrong) ideas are discarded – at least most of the time. When Copernicus insisted that the Earth travelled around the sun, he met some opposition, but nonetheless, within a generation, the paradigm had changed. Imagine such a genetic disposition being overturned in a single generation! It is precisely this ability to improve and to prune ideas that has caused human civilization to advance at such break-neck speed. And if this kind of cultural transmission does occur on another planet, you can be sure that evolution will be swift and effective – just as ours continues to be.

Of course, we don't know what the evolutionary future holds for humanity. It's very difficult to predict where we will be in another million years, because we have no evidence, either in the world around us today or in the fossil record, to show where this rapid cultural Lamarckian evolution will end up. Ironically, science fiction – which it is sometimes easy to dismiss as unrealistic – may give us the best clues. For the past 150 years, people have been thinking up thousands upon thousands of more or less plausible scenarios featuring futuristic humans and future ecosystems. Much of this is, of course, scientifically non-sense, but science fiction authors are among the few who have taken seriously the question of the philosophical implications of a future world – or an alien world – where we have evolved spectacular and new abilities.

Let us do a brief thought experiment. Imagine that you are a member of a highly advanced alien civilization, intent on spreading your legacy throughout the galaxy. You come to an uninhabited planet with the intention of 'seeding' it with some-thing of yourself. What should you put there? One obvious answer is simply to colonize the planet with members of your species – what better way to propagate your biological and cul-tural heritage? At the other extreme, you could place on this planet something that resembled your very distant ancestor, perhaps the Last Universal Common Ancestor of life on your planet, or even just the chemicals that were necessary to get life started on your own world. On Earth, the necessary chemicals were probably RNA, a simpler relative of the familiar DNA. By dropping some RNA into the seas of this barren planet, you might find, in a few billion years, a whole ecosystem had evolved, similar but different to your home planet. Just how similar or how different the ecosystem would be is a topic of fervent argument among biologists.

As we discussed in Chapter Two, some scientists believe that 'rerunning the tape of evolution' would lead to radically different results: no mammals or snails or birds, instead a menagerie of unrecognizable alien-like creatures.* Others believe that although the precise details would be different, we would still recognize the fundamental solutions to common problems – including a two-legged, large-brained, tool-making equivalent of us humans.† In either case, one thing is sure: evolution will act via natural selection. All of the arguments I've made in this book so far would be no less valid whether RNA in the primordial seas had appeared spontaneously or had been placed there by alien visitors. Where the original life-creating molecule comes from doesn't matter – how it develops in the intervening billions of years does.

But there's another possibility. Instead of dropping off a biological molecule, you might be such an advanced alien that you seed this planet with intelligent artificial creatures, specially designed robots that have the capacity to *bypass* natural selection. They are programmed to have the foresight that nature is lacking. Their robotic gazelle descendants would *know* that longer legs are better and would re-engineer their own design to give them longer legs. Similarly, robotic lions would reprogram their own software to enable them to sneak up on prey more stealthily. What would be the end result of such a scenario? Would there still be predators and prey, as I have predicted there must be on alien worlds? Or would these creatures be improving themselves

* See *Wonderful Life* by Stephen J. Gould.
† See *Life's Solution* by Simon Conway Morris.

so rapidly that soon the robot gazelles are building spaceships to escape the robot lions and the robot lions are building supercomputers to design weapons of mass robot-gazelle destruction? This ridiculous scenario is not nearly as trivial as it seems, because the idea touches on some of the most fundamental mechanisms and constraints on evolution. Can intelligence, and the ability to bypass natural selection, also bypass the limits that the natural world imposes? Can we, as an intelligent species, continue to expand and consume, trusting our intelligence to get us out of any impending ecological disasters?

The nature of super artificial intelligence

Imagine that one day an alien civilization – or perhaps our own civilization in the future – invents an artificial intelligence with capabilities greater than its creators. Such a creation would have definite Lamarckian abilities: it could learn from its predecessors, and from its mistakes. It would evolve rapidly and become frighteningly efficient. Some scientists and authors, including Stephen Hawking, have proposed that this is a genuine threat to life on Earth, and possibly to life in the universe. Others (myself included) have more faith that the more super intelligent an organism is, the less inclined it would be to destroy and dominate out of fear. Malicious and dangerous or benign and wise these alien superintelligences may be, but what would the ecosystem look like on a planet that has been inhabited by such artificial organisms for a long period of time? Should we expect something broadly resembling the conclusions I've made in the previous chapters, or something utterly different, breaking all the rules laid down by natural selection?

On the face of it, many of the familiar characteristics of animals and plants that we see around us would simply disappear if the organisms in an ecosystem were all super artificial intelligences. Oxford professor Nick Bostrom, for instance, suggests that a community of artificial intelligences will be sharing information in such an effective and rigid way that many aspects of animal behaviour would be unnecessary.* Large antlers, peacock feathers, colourful flowers, even birdsong – why bother with such peculiar and inefficient ways of communicating simple messages like 'I'm here and I'm strong'? Artificial organisms could achieve the same result simply by sending an email. And if the system is well designed, checks will ensure that the email is a form of honest signalling (see Chapter Eight). No fake Tinder profiles allowed.

Even play would be unnecessary, Bostrom suggests, because organisms will be 'born' (created) already knowing everything they need to know about surviving in the world, without the need to experiment and gain skills. Robotic cheetah cubs will be able to hunt gazelle the day they are switched on, without needing patient practice stalking their litter-mates and playing with smaller animals and injured prey.

A super artificial intelligence presumably has some goal in mind. We may not be able to predict, or even to fathom what such a goal may be. It may be to explore the universe, or to wipe out all other civilizations or, in Bostrom's worst-case scenario, to manufacture paperclips, relentlessly converting all matter in the universe into an ever-expanding pile of . . . paperclips.

If the goal of the originating intelligence is singular – 'manufacture paperclips', 'destroy all other life' – then it is imperative that this message is passed reliably to all of the paperclip-manufacturing robots spreading across the universe. An utterly infallible communication method must be built in. Otherwise the smallest error could cause some of the bots to start manufacturing

---

* Nick Bostrom's book *Superintelligence: Paths, Dangers, Strategies* gives a very detailed analysis of why super artificial intelligence could be so dangerous.

staples instead of paperclips. In that case, competition between the super-intelligent paperclip-manufacturing robots and the staple-manufacturing robots could lead to war and destruction.

How likely is it that this universe of interconnected computers would be doing nothing but communicating, reproducing and carrying out their singular goal? Possibly not very. If such an alien world of artificially intelligent organisms really exists, there are some things it cannot avoid – no matter how intelligent or how well designed. On the one hand, artificial intelligence cannot improve without change, and change brings the risk of mutation. On the other hand, even the cleverest strategy is potentially open to exploitation – game theory cannot be discounted, even by a computer of sci-fi-level superintelligence.

Mutation – a blessing or a curse?

In life (that is, natural organisms like us), mutations occur because the universe has an element of randomness to it. A stray cosmic ray knocks an electron from an atom, and the copying of your DNA is disrupted. An enzyme might be 99.99 per cent specific for a certain protein, but there's always that 0.01 per cent chance that the 'wrong' protein might get sucked in. Errors like these can be disastrous – in general, just one incorrect 'letter' in the encyclopaedia that is your DNA can mean that as an embryo you will be incapable of growing and functioning correctly. Life, then (Earthly or alien), could not have evolved anything like the complexity that it has without some form of error-checking and error-correcting every time that DNA reproduces. By and large, this system works pretty well. The cells in our body are dividing and reproducing all the time – you see that most strikingly as we grow from a 3kg baby to a 70kg adult. Each cell gets a faithful copy of your DNA, and if it doesn't, the result is often cancerous.

However, not all mutations are necessarily this bad. A mutation that causes your neck bones to grow for a bit longer as an embryo may or may not be beneficial to the adult you, but probably won't kill embryo you. Variation in individual organisms inevitably arises through mutations (and also – on Earth at least – through sex).

All of this is very straightforward when considering natural selection, because natural selection has no design and no designer. No one knows in advance which features of an organism are beneficial and which features are harmful. Without that foresight, the *only* way to try out different variants is through random but small mutations and alterations on the theme that already exists.

But what if I *know* what I want? What if I have a plan for how my offspring should look and behave, and I don't want to leave anything to chance?

Consider an artificial intelligence (or even a biological organism) that creates a set of self-replicating intelligent robotic probes to fly off and explore (and colonize) the universe. Each probe will land on a different planet and begin to create new probes like itself, much like Dneprov's *Crabs on the Island*. Will each daughter probe be identical to the parent? Probably not. The parent probe may choose to make them slightly different from one another, with intelligent foresight: one may be optimized for swimming underwater, and one for flying in the air, for example. But will there be any errors, any mutations in the process? It would seem that, like any respectable engineer, the parent probe would make every effort to ensure that each replicated probe is exactly the way it was meant to be. The advantage of evolutionary mutation exists *only* because evolution has no foresight! If you *do* have foresight, it makes sense to dispense with the randomness.

But even if you can be 100 per cent accurate, and are capable of completely stopping bugs from creeping into your software (our hypothetical parent is super intelligent, after all), we have seen that variation is, nonetheless, necessary. Even if you don't want mutations, you need a swimming daughter probe and a flying daughter probe. The offspring of each of those daughters will also change: one granddaughter probe for swimming in deep water and one for shallow water, for instance. As time goes on and the environment on the planet changes, a wide diversity of artificial creatures will arise. Not through the mechanisms we are familiar with on Earth, but diverse nonetheless. Each one will be perfectly engineered to its niche, without the awkward wisdom teeth and appendixes that we carry with us, and that betray our origins as creatures not designed by anyone.

Everything's a game

So does that mean that Bostrom's view of an ecosystem of perfectly prepared individual artificial organisms, fully inter-connected and communicating, will have no need for music or play or art? Would these artificial creatures be, in essence, drones operating for the good of their overall goal (even if just manufacturing paperclips)? Would they be without conflict and without competition? It sounds like a boring alien ecosys-tem, but at least a peaceful one.

Yet even such a well-oiled and super-intelligent community is subject to the laws of mathematics. Game theory, as we dis-cussed in Chapter Seven, is ruthlessly inevitable. If exploitation pays, exploitation will happen. What does this mean for our single-minded superintelligence? A hypothetical paperclip-manufacturing swarm of robots is kept in line by the very programming that gives each organism its purpose. As long as the software is intact, every organism will faithfully carry out its task. But what if a mutation arises that gives a single bot the opportunity to rewrite its code? In fact, we have already seen that the ability to change the programming of your daughter bots is essential. What if, for example, one organism decides that the best way to fulfil its destiny is to cannibalize the other bots? Selfishness is a threat that will be present even in a community of super-intelligent artificial aliens.

In the most perfectly cooperating society of selfless individuals – take the cells in our body as an example – a selfish mutation can be disastrous. When everyone assumes the best intentions in everyone else, the exploiter has a field day. When one of the cells in our body decides to go selfish, reproducing without control and without providing a function for the body, we get cancer.

When everyone is cooperating, allowing open access is the most efficient strategy. In a small village, you don't lock your door when you go out, and in your (healthy) body, your immune system does not identify other cells in your body as invaders. But our swarm of paperclip bots leave their conceptual back doors unlocked, a software vulnerability that is easily exploited. Therefore, a selfish mutation, even if it is not a random mutation but a carefully calculated one, will undermine the integrity of the communal effort. As with all game theory, the success or otherwise of such a mutant depends on the response of the other players. Perhaps the mutant will die out. Perhaps an equilibrium will be reached, where communal bots share a planet with selfish ones. Or perhaps the communal paperclip venture will be doomed.

Now, you might think that, being super intelligent (more intelligent than me, anyway), the alien parent bots will be aware of this danger and will take steps to avoid it. Even our own bodies have cancer detecting and defeating mechanisms, and the human race does not die out just because cancer is always a possibility. However – and here is the catch – we have postulated that this mutation is a rewriting of code in a bot that *is itself super intelligent*. In sci-fi stories like *Terminator* and *The Matrix*, humans make feeble but valiant attempts to fight off an artificial intelligence takeover of planet Earth, attempts that appear doomed (despite the obligatory heroic ending), because our adversary is so much more powerful than us. But in my hypothetical case, the renegade bot has similar capacities to those of its overlord. The odds are now radically different. As J. K. Rowling put it, 'The other side can do magic too.'★ Like any uniform strategy, it seems inevitable that paperclip manufacturing will eventually be halted by alternative strategies that can, at least partially, compete successfully with the orthodoxy. Super intelligent doesn't mean invulnerable.

The Grim Reaper

It might seem that avoiding dying is an essential part of the life of every organism on this planet, from the gazelle that flees the cheetah, to the cheetah that doesn't want to starve; from roses that grow thorns to avoid being eaten, to biochemists who research the cure for cancer. It seems obvious that animals want to live forever – the more time you have on your hands, the more children you can create and that is, after all, the currency of natural selection.

Taking this argument further, we might expect aliens to be smarter and more advanced than us, living for longer and, maybe . . . living forever? We seem tantalizingly close to solving any and every medical problem in the human body. The eccentric millionaire biologist Aubrey de Grey has famously stated that the first human to live to 1,000 has already been born – and he has staked his fortune on building a company to

---

★ *Harry Potter and the Half-Blood Prince*, by J. K. Rowling.

deliver that goal.⋆ Surely alien civilizations are way ahead of that. Surely aliens know how to eliminate death completely?

Science fiction literature is full of suggestions for how we achieve this, for example that we should slowly replace our failing organs with artificial replacements, until we become entirely cyborg – except that we still retain that fundamental essence that is 'us', our 'soul', if you like. The question of whether such a thing is even possible has been debated for centuries, since Descartes first declared mind and body to be separate and irreconcilable.† Many authors, in science fiction as well as scientists like Nick Bostrom, also consider whether we can in fact achieve immortality by uploading ourselves to computer software, thereby completing the transition from biological to artificial organism, while still being 'us'. Even if *we* can't do it, perhaps a sufficiently advanced alien civilization will have developed the technology?

But as an evolutionary biologist, my mind is drawn to a completely different question. Even if we *could* live forever, *would* we? Is it even evolutionarily feasible? Why don't any natural organisms live forever? True, they lack the foresight and the technology of artificial creations, but they have been around for a lot longer. As it turns out, there are good reasons why immortality is a *bad* idea.

Evolutionary theory suggests that death is not a parochial feature of life on Earth. Technologically sophisticated alien civilizations aside, animals on other planets will die. Death is essential for evolution, and – at least until technology can be invented – evolution is the only way that complex animals can arise from simple ones. Death is essential for evolution for three reasons.

⋆ Strategies for Engineered Negligible Senescence, or SENS.
† For the detailed and somewhat bitter argument between modern philosophers on the topic, see David J. Chalmers, *The Conscious Mind: In Search of a Fundamental Theory*, and Daniel C. Dennett, *Consciousness Explained*.

Firstly, and most obviously, if no one died, we would run out of space. Evolution works because organisms have offspring, and because those offspring differ from their parents. If there is no room for children because the parents are refusing to shuffle off the mortal coil, there can be no change to the functions and adaptations of the organisms. Evolution stagnates when the parents don't give way to their children. An immortal amoeba simply would never have evolved eyes.

Secondly, the world is full of changes. No matter how smart we parents are, no matter how much we know (and we really do believe that we *know*) when is the right bedtime for our children, what they should be eating, and where (or even whether) they should go to school, sooner or later, the world will change beneath our feet, and presto: our children are teaching us how to use our smartphones, and how not to embarrass ourselves on Snapchat. In a world full of dinosaurs (literally and figuratively), an asteroid impact would spell disaster if there were no other organisms, with different niches, different solutions to life's problems and different opportunities. On a planet with no death, even a small environmental change could lead to the death of all.

Thirdly, and most importantly, life is full of trade-offs. It is an irreconcilable feature of the universe that you can't have all you want of everything. Immortality comes with costs. If it is, as de Grey suggests, possible to correct the errors in cell replication, to replace faulty organs, and to ensure that our arteries do not clog up, all this comes at a cost in other areas. Perhaps Teflon-coated arteries would never lead to coronary heart disease, but they might reduce your ability to fight infection, or to run up a mountain. An excellent example of evolutionary trade-offs in general is defensive armour. If you are terrified of predators, you can grow a strong shell like a tortoise, and spiky horns like a *Triceratops*, as we saw in Chapter Seven, but you pay for this in agility, losing speed and manoeuvrability. Perhaps it is better to take the chance of being eaten, and move a bit faster? In a world of 200-year-old tortoises, the hare has an advantage, even if she is easily snapped up by a fox. Longevity is a trade-off, and the balance is always in between, never at one extreme.

Imagine a planet on which everyone lived forever. We can speculate what kind of cost they might pay: perhaps like tortoises, they crawl very slowly. What of the individual who can run fast, but doesn't live quite as long, perhaps 1,000 years instead of 1,000,000? Such an individual could be at a huge advantage – being better able to avoid predators and find food, say – so their genes would eventually spread through the population. Evolution shows us, both in evolutionary theory, and in the world we see around us, that making an excessive investment in an extreme stance can't really be justified. Extremists, in any field, are always susceptible to moderates, who can take advantage of the middle ground.

On Earth, many species are very successful through the strategy of living a short but productive lifespan. When mayflies emerge as adults, they live only to reproduce, and quickly die, often after a few hours. To keep things efficient, adult mayflies don't even have working mouths or digestive systems – no time to feed, just mate! Male ants live only to inseminate a queen. Most impressively (by our standards, with our obsession with sex), small shrew-like marsupials, called antechinuses, have such a ferocious mating season that all the males simply die of exhaustion after mating continuously for a couple of weeks. For animals exploiting these niches the idea of living forever would simply be bad judgement. They are well adapted for exploiting their own lifestyle and their own environment, and living a long life doesn't form part of that adaptation. Give a male antechinus immortality and he will pay a cost: his matings will be less frenzied, as he would have to save some energy for mating next year as well. His immune system would not be mortally damaged by constant copulation, but he would be out-competed by his reckless colleagues.

Do the same arguments apply to artificial life? Surely a smart enough alien civilization could design a way around such engineering challenges? Probably not. Trade-offs are too fundamental a feature of the universe. You could use a durable titanium shell for your cyborg skin, or perhaps a less durable but lighter plastic one would do. Yes, you could design batteries that never wear out, but is it really necessary? Perhaps ones that last just 5,000 years would give you the ability to fly faster? Even for a radical, technological, non-Darwinian life form, so many of the principles that we use to understand the evolution of life on Earth still apply, and so it is unlikely that they would design themselves to live forever.

%

On the face of it, artificial organisms seem to open up an unlimited range of possibilities for alien ecosystems. Their ability to redesign themselves, to adapt almost instantly to changes in the environment, and to predict what is needed to help them achieve their goals, means that natural selection simply doesn't apply to them. So can we predict anything at all about what they might be like?

Yes, even if natural selection isn't operating on these creatures, some of the rules of evolution will still apply, no matter how flexible and self-designed they might be. Even super-intelligent artificial life forms are subject to the restrictions imposed on them by game theory – they would, after all, be competing against other super-intelligent organisms like themselves. And some things like mutation, and even death, can't be eliminated just by being incredibly smart.

But are we likely to encounter a planet of artificial life forms? Strangely, there is no sign that the universe has been swarmed by such super-powerful bots. If astrobiologists wonder why we haven't found any indication of alien life so far, we should be doubly confused as to why we haven't found any indication of alien super-life. Once created, surely AI will take over the universe? Well, it hasn't happened yet, so perhaps the risk is less than we thought. Factors such as cooperation and selfishness, and the need to navigate the trade-off between resources and longevity, have prevented bacteria from taking over Earth, and might prevent alien robots from taking over the universe.

And what about artificial life forms that are more modest in their abilities? Could a planet evolve an entire ecosystem all by itself, based on artificial life forms that accelerate evolution by having at least a basic Lamarckian ability, passing their lifetime experiences on to their offspring? It's possible that such organisms may not have such an advantage over natural selection after all, at least not until they can evolve the ability to communicate, cooperate and plan their evolutionary strategies intentionally. If you are an alien species seeding a planet with some prototypical artificial organism, maybe it's better to seed it with Darwinian creatures, rather than Lamarckians.

Is it possible that we, ourselves, are artificial creatures, seeded onto planet Earth by intelligent aliens billions of years ago? The idea that life on Earth began in outer space, known as panspermia, is not a new one, nor is it a ridiculous suggestion in itself. Astronomers Fred Hoyle and Chandra Wickramasinghe suggested that the molecules necessary for life were created elsewhere in the galaxy, and arrived on Earth on meteors. Recent studies strongly suggest that bacterial life is capable of surviving for millions of years in space, so much so that rock ejected from Earth by the asteroid that killed the dinosaurs may already have made its way to the moons of Jupiter – when we finally arrive there, we may well find life based on the same DNA as our own, carried by dormant bacteria on a 60-million-year journey.

But this idea of directed panspermia, that alien intelligence deliberately placed life, or life-giving chemicals, on another planet to jump-start evolution, is not just found in science fiction.* Carl Sagan and Iosif Shklovskii proposed as much in the 1960s.† Perhaps it even happened on Earth. If an alien race deposited the seeds of life on Earth, and then left the planet to its own devices, then that life would evolve according to the same rules as life that arose spontaneously. Most likely, then, we would not be able to distinguish such simple artificial organisms from 'actual' biological ones. Perhaps we are ourselves descended from alien worms deposited on Earth to evolve into whatever may come. If biological life is indistinguishable from artificial life, perhaps it makes no difference. But there is no sign of this, no fingerprint of alien interference. We might as well have evolved naturally – we show all the signs of natural selection pure and simple, with no trace of Lamarckian acceleration.

Unless . . .

We actually *do* have the ability to adapt our evolution, to spread the ideas and experiences of our lifetime both to our own offspring and to others. We have Lamarckian abilities through our culture and through our technology, and what is more, we have the intelligence to know how and when to use them – although it remains to be seen whether we will be intelligent enough to be able to find our way out of the environmental dead end into which our expanding consumption has led us. Using modern techniques of genetic engineering, we can even alter the contents of our genes, eliminate the susceptibility to disease and perhaps even stop ageing. Eventually we may be able to change the shape of our very development, growing an extra arm, or wheels, or whatever takes our fancy. Perhaps our alien seeders knew that

---

* 'The Chase', *Star Trek: The Next Generation*.
† See *Intelligent Life in the Universe* by I. S. Shklovskii and Carl Sagan.

consciousness would evolve. Perhaps this was the grand plan the whole time: Lamarckian artificial organisms could not survive their early evolutionary stages, but would one day mature. Our creators knew this and were patient enough to wait for the day when that would happen. This is possibly an unlikely scenario, but it does leave open the possibility that alien planets will be inhabited by 'artificial' life forms that are nonetheless indistinguishable from what we would expect from natural selection.

11.   Humanity, As We Know It

SPOCK: Captain, we both know that I am not human.
KIRK: Spock, you want to know something? Everybody's human.
SPOCK: I find that remark insulting.

Star Trek VI: The Undiscovered Country

For millennia, philosophers have wrestled with the question: what does it mean to be human? To be fair, until recently it's been a very uneven wrestling match. What it is to be a human is pretty obvious – there's not much mistaking what is a human and what isn't. Humans are distinctive and unique. However, in recent years the waters have become a little muddied. Just a little. New discoveries about our own evolutionary origins and our familial relationships with other human-like species have been grabbing newspaper headlines: just how much did our ancestors interbreed with Neanderthals? As we touched on in the last chapter, genetic engineering has opened ethical cans of worms about altering our own embryonic development and even transferring DNA from one species to another. And as we saw in Chapter Six, our increased understanding of animal cognition – the way that they think, feel and reason – makes us wonder: how genuinely unique are we?

Turning our eyes to space makes us even more uneasy. Of course, people have long speculated about the possibilities of human-like, or even super-human beings beyond Earth. In the ancient world, ideas of angels and demons were considered unremarkable: their existence was obvious, even. More recently, even Johannes Kepler, second only to Copernicus among the

Renaissance astronomers, wrote science fiction about the inhabitants of other planets.*

But if you accept the conclusions that I have presented in this book so far, then you have even more reason to doubt our own uniqueness. If, as I suggest, similar evolutionary processes are taking place on other planets throughout the galaxy, and that life on other planets will be predictable, expected, even possibly *similar* to life on Earth, then what of intelligent, rational alien life? Will it be similar to us? Perhaps *very* similar? Television science fictions such as *Star Trek* portray aliens as basically human, partly to allow the audience to feel rapid empathy for the characters, and partly for budgetary reasons (it's really expensive to make convincing monster costumes). But might it be truer than the writers and producers expected?

Humans are special, no one can doubt that. But if you really believe in the strength and universality of the biological laws that we've been talking about through this book, then this raises a difficult question: maybe we are special, but we are not *unique*. Maybe we belong to a broader category, and any alien that fits into this proposed category would be instantly identifiable to us as having many things in common – very much like the Vulcan Spock from *Star Trek*. Is that so unlikely? What happens if we discover aliens are morphologically similar to us: bilaterally symmetrical, walking upright and manipulating the world with two hands? More pertinently, what if aliens are *cognitively* similar to us: having families and jobs and pets, and speaking a language with a similar structure to ours? Might they be, in some sense – any sense – 'human', as Captain Kirk suggests?

This chapter is less about hard science than the previous ones. My conclusions here are based on the same universal laws of

---

* *Somnium* by Johannes Kepler, 1608.

biology that I have discussed previously, and follow from my own years of research into the ways that animals evolve and behave. Nonetheless, they show my personal worldview – although this worldview is not out of place in a scientific community that has yet to come to a consensus on the topic. I hope that most readers by now agree with my observations and conclusions about how natural selection shapes life on other planets. Now I will apply those conclusions to some philosophical questions, and I don't necessarily expect the readers to agree with my outcomes. However, I'm hoping that your own conclusions, when you have finished this chapter, will be based on the same foundations as mine, even if the outcomes are different.

Personhood

Before trying to solve philosophy's biggest problem – identifying the fundamental nature of humanity – in a single chapter, I want to point out that this is a very different (but related) question to that of personhood. 'Are aliens people?' is not the same question as 'Are aliens human?' The concept of 'person' has a primarily legal significance, as well as philosophical. There are rules and conventions in society about how we treat 'people', and eventually we are going to have to decide how we are going to treat aliens too. Will they have rights under the United Nations 'Universal' Declaration of 'Human' Rights? Or will we be free to exploit them and their resources, much as Europeans exploited the inhabitants of their colonies on Earth? The human history of colonization and exploration didn't generally end well for the explored and colonized.

The question of who gets rights isn't just an issue 'out there', far away from civilization. Throughout history, certain humans have been considered legally 'people', whereas others have been denied that status based on their skin colour, their religion, social status, or even their age. We like to think that we are more kindly these days, but would we really extend this kindness to creatures on other planets? Would they extend it to us? People are genuinely considering the implications of these questions, but without any evidence of what kind of legal system or system of ethics the aliens themselves might have, it is hard to avoid simply projecting existing human law into space. Legal personhood is not a universal property. It depends on culture and history and moral norms, and alien lawyers on other planets will have their own opinions on whether or not humans will be considered 'people'.

We can draw on our ongoing debates on animal rights to throw some light on how we might treat aliens. The question of how we *should* treat non-human animals, and how we actually *do* treat them, is an excellent example of the difficulty of the concept of personhood. Certainly, many campaigners would like to see animals being accorded human rights. Courts have been repeatedly petitioned to grant, for example, chimps the right not to be used in animal experiments, or killer whales not to be kept in captivity, which the petitioners consider 'slavery'. All these petitions have so far been rejected. Society doesn't seem ready to grant legal personhood to animals. For the moment, this is more of a legal definition than a biological or ethical one. But, if animals are not people, how are they different?

Many animal species possess rich complex internal lives that rival those of humans – the evidence for this is no longer in doubt. Many people would be convinced that your dog is genuinely excited to see you when you come home, or that a rat in a laboratory maze is genuinely anxious or scared. Not only do their reactions and behaviours match what we would 'expect' from a sentient being in that situation, but modern technological approaches to neuroscience, like fMRI, show that their mental apparatus – their brains – are similar to ours.★ Sufficiently similar that we are comfortable explaining their emotional reactions using the same mechanisms we use for humans. Some countries, such as Switzerland and Austria, have legislated that animals are not in the legal category of 'objects' – although stopped short of granting them legal personhood. Animals are clearly not just objects – so what are they? Something between object and human – is that half human?

One of the classic tests for human-like personhood is whether an animal has a 'sense of self': an internal perception that they are an individual, different from other animals or objects in the outside world. Sense of self is an important test, because among other things, if you aware of yourself, you have the capacity to suffer. Our legal definition of personhood is at least partly intended to provide a framework to prevent 'people' from suffering.

As we can't ask animals directly whether they feel this way, scientists have used simple tests like the mirror test: do they recognize their reflection in a mirror as being 'them', as opposed to just any other animal – or perhaps they have no recognition of the figure in the mirror at all? By placing a mark on an animal's head where they cannot see it without a mirror, we can observe how they respond when they do look in a mirror. Do they reach up to examine the mark, once they've seen it in reflection? That would imply that they understand that the mirror image is an image of themselves. The mirror test has been used throughout the decades on animals from chimps to dolphins with equivocal results, and remains a very controversial way to determine personhood – there are many reasons why it might not be a

---

★ *What it's Like to be a Dog: And Other Adventures in Animal Neuroscience* by Gregory Berns.

Happy the elephant at Bronx Zoo, examining the mark placed on her head, having seen her reflection in a mirror.

good general test.* Most scientists would not rely exclusively on such an arbitrary experiment to declare general truths about animal mental abilities. But interestingly, one court case that hit the headlines in 2019 involves an Asian elephant, Happy, currently kept on her own at Bronx Zoo in New York. Activists want the court to rule on Happy's legal status as a person, not because elephants in general can pass the mirror test, but because Happy *specifically* was one of the subjects of a classic study of mirror self-recognition in elephants.

In this study, Happy was observed not to show any interest in the mark placed on her head, until she encountered the mirror. Having seen her own reflection, with the mark, she raised her trunk to examine it. This was a convincing demonstration that Happy had a sense that the image she was seeing was 'her'. Animals may or may not know who they are, but this *particular* one does – is *she* a person?

* *Are We Smart Enough to Know How Smart Animals Are?* by Frans de Waal is a good start for anyone wishing to know more about animal cognition and how we measure it.

If the legal status of animals is a minefield, we can anticipate another minefield to cross when it comes to intelligent computers. As artificial intelligence improves, it is inevitable that we will at some point have to ask: is this computer alive? At what point will an artificial intelligence have rights? Is a computer always 'property', or might it achieve, at some point, 'personhood'?* Worse, if it ever becomes possible to upload human consciousness to computers, we would be able to 'kill' millions, if not billions, of personalities by turning off the computer with the flick of a switch.† Is that really the same as killing physical humans? If we can't decide whether or not animals and computers can have some of the rights of humans, we are going to have a tough time deciding what rights we grant to aliens.

These questions about the legal status of personhood aren't going to be resolved easily, and cross a wide range of academic disciplines – a single chapter written by a zoologist such as myself cannot provide all the answers. However, the universals of life – things that must be in common across the universe – can help us with our pressing questions of rights and humanity. Knowing where our status as people comes from and why it evolves seems crucial in weighing up the claims of other potential 'people'. If we know that alien intelligences are likely to be akin to us, and to have evolved their own personhood through similar processes, somehow we will have to come to a consensus about what legal rights accrue together with those processes.

---

* See the excellent episode of *Star Trek: The Next Generation*, 'Measure of a Man'.
† Nick Bostrom deals with this at length in his book *Superintelligence*.

Humanity

Personhood is not universal, because it is so closely tied up with cultural norms – to whom we give respect and rights, and from whom we withhold them. But *humanity* seems to be a more general concept, perhaps more clearly definable than personhood. If something or someone is human, we can see something of ourselves in them. Both 'personhood' and 'humanity' serve to draw lines, and simultaneously blur those lines, between humans and non-humans; so what criteria can we use – rationally and objectively – to shape those lines and boundaries? If the legal status of aliens may or may not be that of a 'person', are there any circumstances under which an alien would be considered 'human'?

There are, essentially, two possible answers to this question. One is the most common-sense retort to applying a 'wider' definition of humanity: *no.* Humans are the only humans. The human species. *Homo sapiens.* Us. That is how humanity has always been defined, and that is what humanity is.

The other possibility is that there is something else, some fundamental property or set of properties that mark an individual as human. So far, from what we have seen on this planet, every human (and it is only true of humans) has this special characteristic. But outside of the Earth, perhaps others will seem to be human as well. If Hollywood producers are right, and the intelligent aliens we encounter look exactly like us (except for some minor make-up additions), are they human?

The species

The common-sense answer falls back on our definition of humans as *Homo sapiens*, and *Homo sapiens* only. Despite the obvious appeal of this straightforward definition, it is extremely problematic. Problematic logically, and also problematic biologically.

The logical problem with defining humans as being *Homo sapiens* is that the definition is essentially circular. We are the only humans that we know, and so to define ourselves in terms of ourselves is of very little use. It adds no information about *what* humans are. We may be able to make distinctions between humans and non-humans, but only if those non-humans are already known not to be human! It is easy to say that a dog is not human, but only because I knew that already. Uniqueness is in general a rubbish basis for a definition. Consider something unique, such as da Vinci's painting the *Mona Lisa*. For sure it isn't van Gogh's *Starry Night*. But is the *real* Mona Lisa different from a cheap postcard of the Mona Lisa? Well, yes, but only in the sense that it's different from *everything else*, because there's only one. This is not a helpful definition of what something is, and saying that humans are just humans is equally unhelpful.

Worse, uniqueness ties us to planet Earth at exactly the point when we need to look for a definition that is valid everywhere. Saying that 'human' means '*Homo sapiens*' can only work on Earth, in the same way that saying 'animal' means 'descended from an opisthokont' (Chapter Three). Yet we can, and do (at least in this book), use the term 'animal' to refer to alien creatures. If we rely on Earth-bound evolutionary relationships to classify life, then we can't use *any* of our common terms for describing aliens, because we don't share an evolutionary history with any of them.

The great Enlightenment philosopher Immanuel Kant struggled with this very problem. All the way back in 1798, he wrote:

> The highest concept of species may be that of a terrestrial rational being, but we will not be able to describe its characteristics because we do not know of a nonterrestrial rational being

which would enable us to refer to its properties and consequently classify that terrestrial being as rational.★

In other words, we think we are human because we are
rational. But how can we know what 'rationality' really means,
unless we have another rational species against which we can
compare ourselves?

The biological problem with the species definition of humanity is, however, even deeper. We are generally confident with the
fact that 'species' exist. Any number of butterfly enthusiasts will
be able to tell you the difference between a small tortoiseshell and
a painted lady (they're not that easy to tell apart). Legions of bird-
watchers will be able to list all the different species that they have
spotted in their lives, or in the past twelve months. A cow is not
a sheep. Even Charles Darwin placed the concept of species right
at the top of his monumental theory of evolution: *The Origin of
Species*.

Yet in a very real sense, species is a problematic concept in biology. Useful, but only to a point. Like all approximations, we need
to know when to use it because it is convenient, and when to discard it as misleading. The usual modern definition of species was
proposed by the evolutionary biologist Ernst Mayr in the 1940s
and talks about a species being a group of organisms that interbreed to produce fertile offspring. Indeed, most species do fit this
definition. Cats breed with cats, but not with dogs. Humans breed
with humans, but not with aliens. We are a separate species.

Still, when you zoom in on many, possibly most, species, the
distinction breaks down. Animals do not suddenly become a new
species at some specific point in time. As populations become more
and more different from each other, on their way to becoming different species, there is always a certain amount of interbreeding

★ *Anthropology from a Pragmatic Point of View* by Immanuel Kant, 1798.

between animals, which you might otherwise think 'belong' to different species. Consider the different species of canids: dogs, grey wolves, red wolves, coyotes and jackals (as well as a few others). They are undoubtedly different species, but they can all breed and produce fertile offspring. Wolves and coyotes look different, act differently and occupy different ecological niches; their behaviour is different and they hunt different prey. They look like they should be different species, but they can and do interbreed.* The same thing is happening today with every species. Dogs and wolves are not separate, but separat*ing*, generation by generation.

Species boundaries are blurry. But in fact, this blurring is expected from the process of evolution. What we see as 'species' only have some categorical value because all the intermediate forms are now extinct. Richard Dawkins explained the process most eloquently:

> The distinction between modern birds, and modern non-birds like mammals, is a clear-cut one only because the intermediates converging backwards on the common ancestor are all dead . . . If we consider all animals that have ever lived instead of just modern animals, such words as 'human' and 'bird' become just as blurred and unclear at the edges as words like 'tall' and 'fat'.†

Think about your own ancestry. You may have been lucky enough to know your grandparents or even your great-grandparents, but what if all your ancestors were still alive, hiding on an island somewhere? You could go from one to the other, child to parent, and never see much of a difference. If the island of living ancestors were large enough, you would

---

* It seems likely that dogs are, in fact, descended from a smaller cousin of the ancestor of modern wolves, rather than from wolf-ancestors directly. See *Dawn of the Dog: The Genesis of a Natural Species* by Janice Koler-Matznick.
† *The Blind Watchmaker* by Richard Dawkins.

eventually encounter your ancestors who have distinctly ape-like characteristics. Far enough back, they would be 'not human'. Your distant ancestors would not be able to mate with modern humans, but quite obviously the generations were sexually compatible at every step along the way.

Defining humans – or aliens – by their formal 'species' is problematic, because species is a problematic concept. Many modern human populations (except those of direct African descent, who are 100 per cent *Homo sapiens*) have as much as 4 per cent of their DNA derived from other species, not just *Homo sapiens*. Neanderthals, and another extinct human species, the Denisovans, seem to have interbred quite freely with *Homo sapiens*. If Neanderthals and Denisovans were alive today, would we consider them human? If yes (as is the current scientific convention), then we cannot define humanity purely as a species definition: humans are a collection of different species. If no, then we ourselves are not purely human, because many of us are a mixture of species. I don't find the species definition of humanity convincing at all.

The case of ancient humans that are not *Homo sapiens* teaches us an important lesson. Sometimes, an evident truth turns out not to be true. How should we adapt? Science is in the business of overthrowing established truths and replacing them with new ones. We think the world is at the centre of the universe, until someone proves that we are in orbit around the Sun. Our perspective of the universe – including our own place in the universe – must change. For countless millennia we have assumed that humans are one species, distinct from all other life on Earth. Now some sophisticated DNA analysis on a tiny bit of fossil bone shows that we are not a monolithic species at all. We are not what we seem – we are a mixture of different species. Our definitions must change, we must adapt. And perhaps we have to be open to adapting our definition of 'human' to fit a universe with multiple human-like intelligent life forms.

Multiple creatures, one humanity

Let us for the moment agree, as I suggest, that it is silly to restrict our definition of human to those animals that have lived on planet Earth for the last few hundreds of thousands of years, and are a mixture of various hominin species such as *sapiens*, Neanderthal and Denisovan. If those other species were alive today, there would be three human species on Earth. Rather similar in appearance, for sure, but distinct nonetheless. Perhaps there would even be four, including the one-metre-high *Homo floresiensis*, the so-called 'hobbit' human, whose fossil remains were discovered on an Indonesian island. *Floresiensis* was strikingly different from us – if they were alive today, it would truly be a world reminiscent of Tolkien's Middle Earth.★

To some, perhaps, the idea of 'humanity' being composed of more than one species sounds outrageous, possibly absurd. But the idea is not fundamentally illogical. One of the most convincing descriptions of such a scenario is by the author C. S. Lewis (of *The Lion, the Witch and the Wardrobe* fame). His science fiction novel *Out of the Silent Planet* describes a human travelling to a world populated by, among other things, three separate, very different, but clearly intelligent species. Most remarkably, all three live in harmony, despite their evident differences in ecology and behaviour. Importantly, all three consider the others (and the Earthling protagonist) to be 'people'.

> Each of them is to the others both what a man is to us and what an animal is to us. They can talk to each other, they can co-operate, they have the same ethics; to that extent a sorn and a

★ *The Lord of the Rings*, by J. R. R. Tolkien.

hross meet like two men. But then each finds the other different, funny, attractive as an animal is attractive.★

Lewis's description sounds idealistic and utopian, and indeed it was most likely influenced by his Christian beliefs, with the alien inhabitants of this planet being a sort of reflection of the idyllic biblical passage describing a magical world where predators don't eat prey: 'The wolf will live with the sheep, and the leopard lie down with the baby goat.'† We call that 'magical', because we cannot draw on any biological principle whereby this would happen. However, Lewis's world is something different and doesn't require magic. These are *rational* creatures, using their power of reason to *decide* that coexistence is preferable. There is nothing more magical about three intelligent alien species living in peace than there is about three nations of Earth living in peace. Even if we don't seem to have managed *that* yet, at least it is, in theory, possible.

The Golden Ticket

One possibility for deciding whether an alien is human or not is that there exists a special trait, an exceptional property, which endows a creature with 'humanity'. A Golden Ticket (as in Roald Dahl's *Charlie and the Chocolate Factory*) that admits them, not to a fantastic world of endless chocolate, but to our club – the human club. Kant thought that this Golden Ticket was rationality – humans are *defined* as rational creatures. He wasn't alone. Saint Augustine, writing 1,000 years earlier, went even further. He thought that:

★ *Out of the Silent Planet*, by C. S. Lewis.
† Isaiah 11:6.

> Whoever is anywhere born a man, that is, a rational, mortal animal, no matter what unusual appearance he presents in colour, movement, sound, nor how peculiar he is in some power, part, or quality of his nature.*

These are extraordinary words. Augustine would have considered many of the science fiction aliens of today to be 'men': Klingons, Borg and Daleks are all rational and mortal, no matter how unusual their appearance, or the planet on which they live!

C. S. Lewis, with his modernizing blend of Christian philosophy, agreed. In a remarkably insightful essay, he wrote of the importance of rationality in identifying our alien equivalents, although he preferred to use the word 'spiritual' instead of 'rational'. Nonetheless, his conclusion was quite clear: physical similarity was not a useful measure of conceptual similarity, 'Those . . . are our real brothers even if they have shells or tusks. It is spiritual, not biological, kinship that counts.'†

Rationality is hard to define. In Chapter Six we looked at the evolution of intelligence, and there seems no doubt that many animals are in many ways what we would call 'rational'. One of the things that seems to be fundamental to rationality is a kind of internal introspection. You think about things before you act. Weigh them up. Should I do this, or should I do that? Scientists and philosophers have vigorously argued over whether animals possess that kind of internal conversation, an internal mental life. Unfortunately, we can't see inside the minds of animals. We can, sort of, see inside the minds of other humans, but only by asking them questions. Without an animal language, we might never know what's really going on inside their heads.

---

* *City of God* by St Augustine. See *Life Concepts from Aristotle to Darwin: On Vegetable Souls* by Lucas John Mix for an excellent discussion of historical and philosophical approaches to the idea of animal and vegetable souls.
† 'Religion and Rocketry' by C. S. Lewis.

Over the ages, others have looked for assorted properties that humans possess and animals do not – a clear-cut dividing line between 'us' and 'them'. One by one, each of these Golden Tickets has fallen by the wayside. Humans were the only animals to have abstract reasoning – until recent and ongoing experiments with crows by Nicola Clayton at Cambridge, and with great apes by Michael Tomasello in Leipzig, showed that many other species are capable of imagining themselves in the viewpoint of others, or imagining themselves at other points in time. Humans were the only animals to have culture – until in the second half of the twentieth century we understood (as we talked about in Chapter Nine) that cultural learning was almost inevitable in complex social groups. Humans were the only animals to make tools – until in 1960 Jane Goodall discovered chimpanzees making termite fishing sticks (Chapter Six), and her professor, Louis Leakey, remarked, 'Now we must redefine tool, redefine Man, or accept chimpanzees as humans.'

The turn of the century naturalist Ernest Thompson Seton, an author of compelling stories of animal behaviour, wrote, 'We and the beasts are kin. Man has nothing that the animals have not at least a vestige of, the animals have nothing that man does not in some degree share.'★

Could there really be no Golden Ticket, after all?

Ticket to talk

Most modern scientists think that there is. And that it is language (Chapter Nine). Language seems to be the only clear trait that distinguishes humans from non-humans. We have language. They don't. True, many species have advanced communication – even

★ *Wild Animals I Have Known* by Ernest Thompson Seton.

very advanced communication – but it doesn't fit the test of true language that we laid out in Chapter Nine. It doesn't give them the possibility to convey a truly unlimited number of concepts; not just that 'there is a leopard over there', but also to ask 'what is the meaning of life?' and 'how can we build a spaceship?' Even Frans de Waal, one of the great champions of the idea that there is a continuous spectrum of cognition between humans and animals, said in 2013, 'If you were to ask what the big difference is, I would say it's probably language.'

But even this enduring distinction between human and non-human is not without difficulties. Why do we insist that Alex the parrot, whom we met in Chapter Six, with the linguistic abilities of a young human child, was not human? If we were to discover that dolphins had a language, would they be human?

The (slightly eccentric) French physician Julien Offray de La Mettrie suggested in 1745 that animals and humans were essentially the same: very complex machines. He even went so far as to suggest that an ape with language would *be* human:

> I have very little doubt that if this animal were properly trained he might at last be taught to pronounce, and consequently to know, a language. Then he would no longer be a wild man, nor a defective man, but he would be a perfect man, a little gentleman, with as much matter or muscle as we have, for thinking and profiting by his education.★

Poor Dr La Mettrie was probably under no illusion that his ideas had any chance of being accepted. In the same year that he published his book suggesting that a talking animal would be human, 35,000 (talking, and very human) slaves were transported from Africa to the colonial countries.† Whether or not

---

★ *Man a Machine* by Julien Offray de La Mettrie.
† https://www.slavevoyages.org/.

there is a single trait, a Golden Ticket to being human, if we can't recognize humanity in humans, it is not at all clear that we humans would be prepared to recognize that humanizing trait in an Earth animal or in an alien.

Certainly any aliens who have the technology to come visit us on Earth must also have a language – are they automatically human? Setting up a single trait, like language, as the hurdle over which a creature must jump to become human does not seem to solve our dilemma. The Golden Ticket can be used – and almost certainly *will* be used – to help define personhood. Aliens that can talk are clearly to be afforded legal rights that alien bacteria do not possess. Perhaps one day animals will be afforded rights based on their linguistic capabilities. But we still don't seem to be any nearer to understanding the universal nature – if there is such a thing – of humanity.

Are we on the right track?

It is, of course, possible that this is a pointless exercise. Perhaps there is no universal 'type' of humanity. Perhaps what is special about humans is quite parochial to Earth, and aliens on another planet will share nothing in common with us. They may have language and technology, but nothing that would allow us to say, 'We see ourselves in them.' Intelligent aliens may simply be too different from us to countenance the possibility of giving them the label 'human'.

They may be very different physically; aliens that do not have a discrete body as we do, for example, might have a form of intelligence and cognition that we simply don't recognize. However, I have argued throughout this book that while such bizarrely unconventional life forms might be *possible*, surely more familiar animal-like creatures are *much* more likely.

It is also possible that aliens might be very different mentally. In Chapter Five we talked about electric fishes and how their perception, and therefore their mental images of the world, must be radically different from ours. Could they ever share enough with us to be considered human, even if they had language?

Are whales really fish?

In Chapter Three, I mentioned that the claim in *Moby-Dick* that whales are obviously fish, and not mammals, is a claim that cannot be dismissed out of hand, despite the fact that whales are mammals by virtue of their evolutionary heritage. Evolutionary heritage seems to have a special status as a method by which to classify the organisms of Earth. As Richard Dawkins put it, there are many different ways to classify books in a library, none objectively better than the other. But, he claims, there is only one 'correct' way to classify life objectively – according to its family tree.* This was a revolutionary idea that has held sway for the past 150 years, ever since Darwin pointed out that all life on Earth is related to each other.

But, patently, the family relationship tree cannot help us to classify aliens; at least, it cannot help us to classify aliens and humans in the same tree. Are humans more related to Vulcans than to Klingons (if they existed)? We would not be related to either. Or are we more related to bacteria on Mars than to intelligent spacefaring aliens from another solar system? When creatures share no common ancestor at all, 'how related' they are is not something that can be measured.

If, as it would seem from everything we've discussed so far, there are common processes at work on life across the universe,

* *The Blind Watchmaker* by Richard Dawkins.

then perhaps these processes will lead to similar solutions. Even if our genetics have nothing in common with the equivalent system in aliens, perhaps our similarity can be measured by the evolutionary *processes* we share in common. If two creatures on different planets fill the same niche, solve the same survival and reproduction problems using the same solutions, isn't it churlish to say, 'No, these cannot be related because they don't share a common ancestor'? We may simply need to redefine the word 'related'.

If similar processes lead to convergent evolution on different planets, leading to 'human-like' species across the galaxy, can we identify what it is that makes them 'human-like'? If there is some such property – not like a Golden Ticket, but perhaps a suite of properties – would this be similar to what Kirk meant when he said, 'Everybody's human'?

The human condition

We actually have a term for this human-like syndrome here on Earth. We call it the 'human condition'. The *Oxford English Dictionary* defines 'human condition' unhelpfully as:

> The state or condition of being human, (also) the condition of human beings collectively.

However, it also notes that the human condition means:

> *esp. regarded as being inherently problematic or flawed.*

While not particularly scientific, this latter part of the definition could be helpful. We explore the human condition through art, literature, music and dance. It is something that we all know about internally, but does not seem to lend itself to a concrete external definition. Nonetheless, you would generally agree if I said, 'Shakespeare is skilled at describing the human condition.'

His characters such as Macbeth, King Lear and, especially, Hamlet display not just our outward properties, skills and achievements, but mostly our flaws: jealousy, greed, doubt, regret, mercy – and the lack of it. Perhaps aliens that share these traits would be very recognizable, almost human. In the same *Star Trek* film that supplied the quote which opened this chapter, one of the alien characters declares, 'You have not experienced Shakespeare until you have read him in the original Klingon.'

It does seem to be empirically true that all the human cultures on Earth have certain things in common. Research on cross-cultural variation shows that certain behaviours and practices are found in too many different cultures to be mere coincidence: decorative art, family feasting, funeral rites, inheritance rules, etc.* Professor of anthropology Donald Brown compiled a list of hundreds of such shared practices in cultures around the world.† Why do we see such similarity between cultures? It may seem that I share very little in common with the day-to-day life of a person my age in a hunter-gatherer society. Yet many of his traditions and customs would be instantly familiar to me. He shares jokes with his friends, tells stories, has a sense of self-responsibility and consequent self-control. He gossips, wonders about the meaning of his dreams and speaks to young children in baby talk. How can so many behaviours be shared between different groups of people?

Of course, one answer is that these behaviours are directly determined genetically. Despite what seem to us to be obvious physical differences between different populations, humans are incredibly similar to each other genetically. If we share 98 per cent of our genome with chimpanzees, we share about

---

* https://hraf.yale.edu/.
† I've taken this list from *The Blank Slate: The Modern Denial of Human Nature* by Steven Pinker, but he derives the list from *Human Universals* by Donald Brown.

99.95 per cent with other humans. Perhaps surprisingly, I am as genetically similar to a Siberian Yupik as I am to the lecturer in Early Modern British History in the office next door to me (I haven't actually done this DNA test, but the claim is likely to be true, based on large-scale statistical studies).* So perhaps all the similarity we see in behaviour across all human societies is simply due to the fact that all humans are, well, genetically human?

Claims like this make me a little uneasy, as they touch on the decidedly controversial field of sociobiology, which claims to explain almost all of human behaviour by our evolutionary heritage.† Behaviour is a very complicated and multi-factored phenomenon, and we (i.e. scientists) are not yet at the level of sophistication where we can confidently explain how a set of genes can on their own lead to complex behaviour like 'dream interpretation'.

Studying animals in the wild, we see a great deal of variation in behaviour, some of which is genetic and some that is culturally transmitted. Some birds copy the song of their neighbours, with a certain amount of their own personal innovation, and their song is further copied by their own neighbours, and so on. In the end, over a long distance, the song is almost unrecognizable. As we saw in Chapter Ten, birdsong 'dialects' are produced

---

* In particular, some studies show that Europeans are more closely related to Asians than they are to other Europeans! If you really want to delve into this murky field, try 'Deconstructing the relationship between genetics and race', by Bamshad, Wooding, Salisbury and Stephens, in *Nature Reviews Genetics* (2004).

† The doyen of human sociobiology, Desmond Morris, wrote popular books such as *The Naked Ape: A Zoologist's Study of the Human Animal* and *Manwatching: A Field Guide to Human Behaviour*. Whatever you think of his ideas, if you want to learn more, those are at least places to start. For a more rigorous read, try E. O. Wilson's book *Sociobiology: The New Synthesis*.

by imperfect copying. This is a widespread phenomenon, not restricted to birds – I studied a similar effect that generated dialects in the songs of hyraxes. So while genes may define the fact that a bird or a hyrax sings at all, and also what kind of sounds they make, the precise song is such a complex interplay between biology and environment, it's almost inconceivable that the subtle differences in song can be just down to genetics.

So if similarities in human behaviour aren't simply due to the fact that humans are genetically similar, what could produce them?

Barring any even less likely explanations, such as a divinely dictated rule book, the obvious explanation is that these common human behaviours – etiquette, hairstyles, tabooed food – are simply a convergent and effective way for a group of animals to live together in a society as complex and as challenging as human society. We *need* to display body adornment and hospitality. These are natural, almost inevitable, consequences of our being human – and that is despite the huge differences in social structure between a Cambridge college and a Siberian hunter-gatherer village!

The processes involved in the evolution of a social animal group (Chapter Seven) are likely shared across the universe. Would it be particularly surprising if an alien society that reached a level of technology similar to ours would have also developed many of the same common traits found among diverse human populations? Brown's list of shared cultural practices, while not definitive or absolute, shows a lot of behaviours that appear to be adaptively advantageous. 'Hygienic care' seems important in any large society, and 'gift giving' as a way of strengthening social bonds does not seem like it should be peculiar to Earth. If aliens were to evolve many of this range of behaviours common to humans, then they would share much of 'the human condition'. *Hamlet* may be as appealing to them as it is to us.

What is war good for?

War. What is it good for? According to 1970s hit single by the aptly named Edwin Starr: absolutely nothing. Or perhaps not. Some researchers, such as Peter Turchin, have argued that war was absolutely fundamental in the evolution of human society as it is today.* For one thing, the invention of weapons that can kill at a distance – spears, then bows and arrows, then guns and missiles – lend a kind of un-biological equality to human society, and therefore give advantage to brain over brawn. A subordinate gorilla can fight the dominant for his position, and risk being badly beaten, or worse. Alternatively, if he can just shoot the dominant dead, physical prowess becomes less of a predictor of success.

Must alien civilizations have followed the same path? If we are looking for human traits to recognize in aliens, the warlike trait is one that we'd rather not find – for our own sakes. But perhaps warring is inevitable in any civilization that grows to achieve technological abilities that would allow them to travel between the stars. Whether or not conflict *must* be necessary for the innovations we have seen in human society is not a question for zoologists like myself. We need to be wary of reading too much into our own history.

What is more, our perspective on conflict is biased by our observations of social behaviour in primates. Most large primates live in societies where males compete strongly for females. A dominant male gorilla keeps females in a harem and will fight a rival male who wants to mate with them. Chimpanzees live in groups with multiple males and females, but they also have a

* *Ultrasociety: How 10,000 Years of War Made Humans the Greatest Cooperators on Earth* by Peter Turchin.

dominance hierarchy which can be the basis of vicious competition for those who want to move up the ladder.

This violence arises from the way that primate society forms around the twin resources of food and mates. Surprisingly both gorillas and chimpanzees live in such violent societies largely because their food resources (leaves for gorillas and mostly fruit for chimpanzees) are relatively *abundant*, not because they are scarce. Abundant food means that, to be successful relative to others, males must monopolize females. Everyone has enough to eat – but if you can mate more than others then you'll have more of your genes in the next generation.

Many smaller primates have far less male–male violence, because their food is hard to find, spread out and not easily shared. Why be a bully if, most of the time, no one even has any lunch money to hand over to you?

So while violence is inherent in human society, even in other Earthling societies as well, we don't really have a good theory for how violence might evolve and adapt on alien worlds. It is quite possible that on other planets concepts like 'male' and 'female' may not exist. Certainly, in this book I have avoided drawing too many conclusions based on Earthling sex, because without understanding the alien equivalent of DNA, we can't know how familiar processes like 'sibling rivalry' might play out on another planet. Perhaps the underlying 'alien nature' on other planets is not for strong competition between members of a group. We cannot yet say whether or not we should arm ourselves with space weapons in anticipation of contact with alien species.

Still, Turchin has a point. Those primate species displaying less aggression usually organize into smaller groups, often just a monogamous mating pair and their offspring. When crucial resources such as food or shelter are few and far between, and are easily monopolized by a single family, large social groups are unlikely to provide an advantage. Small family groups like this may never evolve into large technological societies. In contrast, a large society – unless the society is made up of clones like bees or ants – means a lot of conflict of interest between the individuals. Aggression and competition would be inevitable and in turn would drive innovation and new ways to be aggressive and to compete. The depressing truth is that violence may be necessary for large-scale cooperation and innovation after all.

We, as human society, still retain much of the violence that was necessary, it seems, to make us what we are today. Any alien species would probably have cause to fear us, as much as vice versa. But, as part of our 'human condition', we also appreciate the paradox in our warlike past (and present). If an alien could resonate with the words of *Henry V*, it seems to me that we might have more to see in common with them than a need to fear them.

> We few, we happy few, we band of brothers;
> For he to-day that sheds his blood with me
> Shall be my brother . . .*

A universal human creation myth

We have a pretty good idea of how humans evolved on Earth, although many of the most important details are still missing. We particularly want to know how language arose, and how and when self-awareness came about in humans. But all the same, there does seem to have been an evolutionary pathway that involved increasing cognitive complexity, tied to increased human sociality. At some point, our ancestors seem to have hit a critical mass of smartness, and our species became a runaway 'success', with each new technological and social development

* *Henry V*, Act IV, Scene 3, by William Shakespeare.

fuelling yet more 'progress' – eventually ending up with the internet and cat videos.

Some of human evolutionary history is clearly very closely tied to the conditions that existed on Earth at the particular times of particular innovations. Our adaptations to be bipedal, walking on two feet (as opposed to four, like most primates), was clearly absolutely crucial to humanity – for one thing, it freed up our hands to be used for manipulation and for making tools. But this may have been a fluke of Earth's history. Exactly why humans became bipedal is still hotly debated, but many theories revolve around a shift from living in trees to living on grasslands. Perhaps when our ancestors began to live on the ground there may have been an advantage to standing tall to see potential prey and predators over the long grass. Or perhaps our ability to carry food in our arms was the key innovation, or perhaps walking on two legs kept us cool, by exposing only the top of our heads to the sun. Most likely, it was a combination of all of them. However, it would be reckless to suggest that the evolution of intelligent aliens would necessarily be driven by the same process. Clearly, it is unlikely that alien grass arrived to replace alien trees at exactly the right time to push an already intelligent species into a tool-making habit. This was a very Earth-specific transition.

However, there may yet exist a very general story of the evolution of creatures like us, which is shared on many different planets, without reference to the specific ecological or physical conditions on those planets. The pioneering evolutionary biologist John Maynard Smith, along with the theoretician Eörs Szathmáry, wrote an influential book describing what they felt were the most important innovations during the evolution of life on Earth.* Although all of these innovations were absolutely

---

* *The Major Transitions in Evolution* by John Maynard Smith and Eörs Szathmáry.

essential for life as we know it, many of them (such as DNA and sex) can't really be assumed to have occurred on other planets. But the principle of spotting the most important evolutionary innovations is a very useful one. Can we do the same thing, while remaining neutral about the specific details of alien life?

Perhaps if there is a universal story describing the evolution of humans wherever they live, it might go something like this:

Early life was simple, gaining energy from non-living sources; perhaps mostly from the star around which the planet orbits, but also directly from the heat of the planet, and maybe even other sources, like radiation.

The first innovation was that some life forms (I'll call them 'predators') began to get their energy from others ('prey'), *exploiting* the work of others in harnessing energy from nature (Chapter Three). Freeloading is always an option, and game theory would seem to indicate that the evolution of this kind of 'cheating' is inevitable.

Both predators and prey are competing to achieve their goals of eating, and avoiding being eaten. *Movement* would then evolve (Chapter Four).

Once organisms can move, *social behaviour* follows (Chapter Seven). Prey animals can reduce their chances of being eaten by aggregating, and this opens the possibility of more active defensive strategies: sentinel behaviour, building structures, etc.

If any two organisms are to associate together, *communication* is necessary (Chapter Five), at the very least so that they can find each other.

At this point (if not before), the complex interactions between organisms, both those that are helping each other and those that are competing (either with similar organisms, or with predators/prey), lead to the evolution of *intelligence* (Chapter Six): the ability to predict the world and to make decisions that are beneficial to you.

The combination of communication, social behaviour and intelligence leads to the evolution of communication systems that can contain large amounts of *information* (Chapter Eight), leading to an ecosystem that would be very familiar to us. Alien creatures will be singing like birds, roaring like lions and whistling like dolphins, even if their precise forms, and even the chemical makeup of their bodies, will be entirely unexpected.

How long such an ecosystem continues like this, we don't know. Perhaps the next step is incredibly unlikely. We know that it occurred at least once in the universe, but it took at least 3 billion years from the first step in our story. Whatever the reasons and whatever the mechanism, at some point, complex communication evolves into *language* (Chapter Nine).

Finally, possibly inevitably, a social and intelligent organism, with the skill of language, develops complex technology. It is hard to see how any other outcome is possible. Soon, they will be building spaceships and exploring the universe – if they manage to avoid destroying themselves first.

This sequence of evolutionary events would appear to be something close to the sequence of events on Earth that led to the evolution of humans. If the same sequence occurred on another planet, leading to the evolution of similarly social, intelligent, linguistic and technological organisms, would we really deny them the label of 'human'?

12.   Epilogue

I wonder to what extent my description of aliens has been convincing. I suspect that a few people may take issue with some of my assumptions – there are *always* assumptions – and will have come to different conclusions about the way that aliens live, the way they behave, and possibly their intentions towards us, benign or otherwise. But as long as you have drawn *some* conclusions – they don't have to be the same as mine – I consider this a success. My main goal is to convince you that we *can* know what aliens are like. Some people claim that there is a lack of data, and that this makes speculating about alien life pointless. This isn't true. We are overflowing with data on life, and it doesn't matter in the end that this life is here on Earth, and not on Mars. Life follows rules. Understanding those rules leads us to understand life everywhere.

I know that you wanted me to tell you what aliens look like. Most likely you wanted to know whether or not they are green. I suspect that no small number of readers wanted to know whether they have sex, and whether we would ever be able to have sex with them. What we want to know about aliens is to a large extent driven by our sci-fi representations of aliens in film and television. Human–alien hybrids are common in science fiction. But conversely, our fictional representations are to some extent driven by what we want to know. Good science fiction explores the most perplexing of questions by removing any barriers of 'sensible reality'. *Star Trek: The Next Generation* – which I consider the Shakespeare of science fiction – is an excellent example of this. So the questions about aliens that we cannot yet answer are still really good questions.

But I believe that I've done something better. The chances of us encountering intelligent aliens is so remote as to be almost dismissed. Even if we were to receive (and respond to) a message from an alien civilization, the chances are that they would be so distant from us – tens if not hundreds of light years away – that we are unlikely to get a response in our own lifetimes. I do not believe that I will ever sit on a rocky alien hillside, watching through my binoculars as the alien equivalents of wolf cubs frolic outside their den. Perhaps the only way we can really study alien life is through evolutionary theory. Understanding the constraints on life and applying them to the physical conditions of another planet, we can come as close as is likely to be possible to being 'alien zoologists'. But scientists on Earth continue to puzzle over how to understand alien life, even if we know we are unlikely to see it ourselves. We look for algorithms to detect and translate alien signals, but we also look at those behaviours of animals on Earth that we expect to be similar to those on other planets. Understanding how and why animals cooperate, communicate and solve problems brings us a deeper understanding of aliens even if we can't study them in person.

You might think this book was just about aliens, but it was really also about life in general: *all* life, life in its most fundamental sense. It was about life on Earth as much as about life on other planets. Not a catalogue of life on one planet or another, but about understanding what life is, why it is, and what it has in common with all other life. So many of the excellent natural history programmes on television are about showing us the diversity of life, but remarkably little about the unifying features of life. This is not too surprising, as most of us are still unaware – even in these days of the great David Attenborough – of so much of the diversity of the catalogue of life on Earth. First we must describe, then understand.

To be honest, many of the concepts I've touched on in this

book are genuinely complex concepts. I have tried to simplify
ideas of kin selection and game theory, for instance, leaving out
a lot of the complex details, despite the fact that many academ-
ics will be deeply distressed at my cavalier approach. I believe
this is just fine. Just as life on other planets will conform to general rules, if not to the specifics of the physical constraints of
Earth, so the rules themselves will be more applicable in a general sense. Even if we omit the mathematical fundamentals of
why evolution works the way it does, the conclusions I've drawn
about the nature of alien life will not be substantially harmed. In
Chapter One, I talked about complex systems being essentially
impossible to predict. This is true, but they are also robust to the
approximations we draw. Sometimes, searching for an approxi-
mate solution is more accurate than seeking a precise one.

If we can extend the rules of biology on Earth to apply on other
planets, then we can conclude all the more confidently that we
are similar to other life on Earth – and not just because we share
a common ancestor. The rules of biology are acting on us all the
time. The twentieth-century father of the study of animal behaviour, Niko Tinbergen, proposed that an explanation of any
animal behaviour must be explained in four different ways. Two
of the explanations concern mechanism: how does it work, and
how did it develop (in the body)? The other two explanations
concern *why*: why did evolutionary history lead to a behaviour,
and what is the evolutionary advantage that it provides?

For example, a wolf has sharp teeth because of the strong cal-
cification that makes the teeth hard. We can also explain a wolf's
teeth by looking at the way the embryo develops, with the outer
layer of the embryo's cells thickening, and later becoming calci-
fied. Both of these are mechanistic explanations. We can also ask
'why' questions to explain wolf teeth. Wolves have sharp teeth
because they come from a long line of predatory mammals,
inheriting their body plan and body parts from their ancestors.

The carnivorous ancestors of wolves stretch far back in time, nearly 80 million years, when the wolf ancestor split from creatures destined to become modern (toothless) pangolins – long before the dinosaurs became extinct. Of course wolves have sharp teeth: they are the descendants of ancient sharp-toothed creatures.

But none of these three explanations are particularly relevant to aliens. On an alien planet, calcium may not be the mineral used for strong teeth. Alien embryos will almost certainly not develop in the same pattern as Earthling embryos. And for sure alien wolves will not share a common ancestry with our wolves here. But Tinbergen's fourth explanation is the one that most people turn to instinctively. Why do wolves have sharp teeth? All the better to eat you with! This fourth explanation – what scientists call the 'ultimate' (as in long-term) explanation, and Aristotle would call the 'final cause' – holds true just as much on other worlds as it does on Earth.

So what does all this mean for the future of space exploration, the discovery of intelligent aliens, and even for the continued coexistence of humans on this planet with other intelligent but non-linguistic species? When First Contact actually happens, of some things we can be sure, and many others will be a complete surprise. One way to prepare ourselves mentally and practically for First Contact is to understand these similarities, and to reconcile ourselves to the fact that there are certain properties that intelligent life *must* have. We might even find it difficult to recognize that an alien species is intelligent. But if that intelligence is directed towards solving similar problems as ours, similar 'ultimate explanations' for how that intelligence evolved will apply. We will already have something in common. Aliens may be very different sizes and shapes from us, but their behaviour, how they move and feed and come together in societies, will be similar to ours. In the introduction to this book I said, 'Who cares if they are green or blue, as long as we both have families and pets, read and write books, and care for our children and our relatives?' I hope by this point you agree with me that this situation is actually very likely. Those evolutionary forces that push us to be the way we are must also be pushing life on other planets to be like us.

At the same time, we must come to terms with the potential diversity of intelligent life, and our neighbours on this planet provide us with a useful testing ground for these ideas. Animals may have evolved under the same physical conditions as us, but all species have taken different evolutionary trajectories, and arrived at a staggeringly diverse set of solutions to the problem of surviving. Not 'lesser humans', they have evolved to suit their environment without the need for aeroplanes and television and, above all, without the need for language. Language unites all humans, even if it seems sometimes to divide us. In reality, we recognize of all humans that their language ability gives us a window into their mind, even if the details of that language are confusing to us. That in itself tells us a lot about why humans are different, and what we can expect from life across the universe.

How can you prepare yourself for the day we discover aliens? Astrobiology is a tiny field of enquiry at the moment, but it is growing. You will find relatively little scientific writing on the subject, although I have pointed to some very useful sources at the end of this book. Really good science fiction, like Fred Hoyle's *The Black Cloud*, can effectively challenge the somewhat trivial and unthinking representation of aliens in the mass media. Ironically, the older science fiction is, the less likely it is to have been tainted by modern preconceptions about alien life, and therefore the more accurate it may be. But most of all, take out your binoculars and look at the little aliens we have here on Earth. From foxes foraging in cities to Arctic terns soaring overhead on their way to the South Pole, these life forms encapsulate such a vast range of possibilities that they must share at least some features with the inhabitants of other planets.

Ultimately, though, our relationship with alien species – even if it must be a relationship that exists only in our minds – will be coloured largely by how we see ourselves in direct comparison, or perhaps competition, with them. Will we be threatened by them? Or will we perhaps threaten them? Are they smarter, stronger, more warlike or more docile? Even if 'colonization' is not a practical option, our sense of who we are and why we are here will be put on a completely different footing should we discover that we are not the only intelligent species in the universe.

On the other hand, a direct biological comparison between us and alien intelligences could bring us to a more complete and more satisfying description of life in the universe. When we encounter our neighbours in the galaxy, we will be struck by how different from us they are. But if we look at Tinbergen's ultimate explanation, Aristotle's final cause, perhaps we will be able to come to an acceptance of a wider type of humanity; one that has room for creatures that don't look like us, no matter what planet they inhabit.

Acknowledgements

For scientists, writing a first book aimed at a general audience is kind of like having a first baby. We really don't know what to do, we're convinced that whatever we're doing is wrong, and we're going to end up with a delinquent, sullen teenager. Like having a baby, you also get up in the middle of the night, having suddenly realized that there's something important you've forgotten, and need to do immediately. Lots of help and advice along the way eases those worries somewhat, even if it doesn't help you sleep at night.

My wife and children have more than bravely put up with my unconventional writing activities. I especially want to thank my son, Simon, and my father, Lester, for reading each and every chapter as it was written and providing detailed and honest (often too honest) comments. My volunteer beta readers, Jordan Habiby from Knoxville, Tennessee, Holly Root-Gutteridge at the University of Sussex, Morgan Gustison at the University of Texas, Alecia Carter at University College London, and Emma Weisblatt here in Cambridge, bravely took up the challenge of getting through the whole manuscript in two weeks, and provided fantastic feedback.

The inspiration for this book draws on many different experiences in many parts of the world. Sara Waller, the irrepressible philosopher and researcher in animal cognition, not only gave crucial feedback on many of the chapters, but also single-handedly organized our wolf research site in Yellowstone National Park for many years. Together with dog-whisperer Jessica Owens and hyena advocate Amy Clare Fontaine (a fellow author and source of encouragement), we spent many a day and night battling through deep snow in temperatures at which plastic tie-wraps snap like biscuits. All to find out what the wolves are saying. Other animal communication scientists who have helped me build up to the point where I could write this book include wolf biologist Holly Root-Gutteridge, chickadee experts Carrie Branch at Cornell and Todd Freeberg at the University of Tennessee, hyraxer Amiyaal Ilany at Bar Ilan University, and marine biologist Nadav Sashar at Ben Gurion University. Dan Blumstein at UCLA has been my long-term mentor and source of encouragement, without whom I would probably never even have started writing. Two more scientists who deserve mention are the unmistakeable beanie-hatted physicist and SETI advocate Laurance Doyle, who started me on this journey of linking animals and aliens, and Simon Conway Morris, another larger than life personality, who has given me useful input on and off for the last thirty-five years, since he unlocked the University's geology museum late at night, to give nineteen-year-old me, and two other undergraduates, a lesson on interpreting dinosaur bones.

Venturing into a new field on your own is hazardous (I know this from years of working in the wild), and you need someone with experience to show you the way. My literary agent, Michael Alcock, believed in this project, and in my ability to carry it out, and paved the way to get it off the ground. Daniel Crewe and Connor Brown at Viking (Penguin) also had an enthusiasm that kept me going, without hesitating to keep me on the straight and narrow, and Katherine Ailes did the copy-editing with insight and intuition, and also had some great ideas for improvement.

Girton College has been a fantastic place to write, with a beautiful office offering sweeping views over the manicured grounds. Plus, I want to thank my undergraduate students, because teaching has, among other things, made me realize how much I like talking for hours and hours about this stuff.

Finally, to my dog, Darwin. We have walked well over 20,000km together in the last twelve years, and even if we've slowed down recently, those quiet, pensive hours have been among my most productive.

Dr Arik Kershenbaum
Girton College, Cambridge
July 2019

Further Reading

1. Introduction

THE PLANET FACTORY: EXOPLANETS AND THE SEARCH FOR A
SECOND EARTH BY ELIZABETH TASKER
A clear and easy-reading guide to the science of exoplanets, how they
are found, and what they are like.

INTELLIGENT LIFE IN THE UNIVERSE BY
I. S. SHKLOVSKII AND CARL SAGAN
Sagan's quirky and appealing style gives a rather sixties feel to this
discussion of the nature of extraterrestrial life.

THE SELFISH GENE BY RICHARD DAWKINS
Probably the most famous popular-science book in the field of evolu-
tionary biology, and rightly so. Crystal clear and to the point, *The
Selfish Gene* is essential reading.

THE BLIND WATCHMAKER BY RICHARD DAWKINS
There are few books that are as essential for understanding the nature
of natural selection as Richard Dawkins' *Blind Watchmaker*. Even
more comprehensive than *The Selfish Gene*, this book is probably the
first to read if you want to understand the nature of life in the
universe.

DARWIN'S DANGEROUS IDEA: EVOLUTION AND THE MEANINGS OF LIFE BY DANIEL DENNETT

Dennett presents the ideas of natural selection from the position of a rigorous philosopher. Worth reading for a more in-depth understanding, but not particularly easy reading.

ARE THE PLANETS INHABITED? BY EDWARD WALTER MAUNDER

A charming and well-written short book by the turn of the century astronomer, who made clear, scientific speculations about the possibility of life on other planets. Freely available online.

THE IMPACT OF DISCOVERING LIFE BEYOND EARTH EDITED BY STEVEN J. DICK

A collection of essays by prominent scientists in the Search for Extra Terrestrial Intelligence (SETI).

2. Form vs Function: What is Common Across Worlds?

'POSSIBLE WORLDS' BY J. B. S. HALDANE

J. B. S. Haldane is one of the most entertaining and lucid writers about biology. Very much an 'old-school' biologist (1892–1964), his humour and insight make his essays well worth reading.

THE BLACK CLOUD BY FRED HOYLE

Possibly the greatest hard science fiction novel of all time, written by one of the twentieth century's greatest astronomers. The story tells of the discovery of alien intelligence inside an interstellar cloud of gas. However, the genius of the book is in revealing the processes by which scientists work together to understand the unknown.

RESTLESS CREATURES: THE STORY OF LIFE IN TEN MOVEMENTS BY MATT WILKINSON

Matt Wilkinson is an expert on pterosaur flight, but his book on animal movement in general is a good read, and will be particularly useful for Chapter Four.

LIFE'S SOLUTION: INEVITABLE HUMANS IN A LONELY UNIVERSE BY SIMON CONWAY MORRIS

This comprehensive and slightly technical book is the standard repository of evidence for convergent evolution in just about every trait of life that we observe on Earth. Conway Morris argues powerfully that this convergence is a universal property, and these arguments have been instrumental in the crafting of the ideas I present here.

WONDERFUL LIFE: THE BURGESS SHALE AND THE NATURE OF HISTORY BY STEPHEN J. GOULD

In direct contrast to Conway Morris's suggestions that almost all traits are convergent, Stephen J. Gould (an associate of Conway Morris) argues that the outcome of evolution is not to be predicted. You are welcome to read both books and make your own conclusions.

WHEN LIFE NEARLY DIED: THE GREATEST MASS EXTINCTION OF ALL TIME BY MICHAEL J. BENTON

This book makes slightly depressing reading, given the current environmental change we are experiencing in the world now. However, for that reason, understanding this little-known period of Earth's history is even more important.

THE BLIND WATCHMAKER BY RICHARD DAWKINS

See Chapter One.

THE RED QUEEN: SEX AND THE EVOLUTION OF
HUMAN NATURE BY MATT RIDLEY

A slightly heavy but nonetheless essential look at the true nature of sex, i.e. why it evolved, not how it is done. Although it's hard to say whether sex exists on other planets, having read this book, you might decide that I am wrong, and sex is utterly inevitable.

ASTROBIOLOGY: UNDERSTANDING LIFE IN
THE UNIVERSE BY CHARLES S. COCKELL

A good general and accessible textbook on astrobiology. Set at a low undergraduate level, it is nonetheless totally readable for anyone interested in more of the technical details of the topic.

LIFE IN SPACE: ASTROBIOLOGY FOR
EVERYONE BY LUCAS JOHN MIX

Not a textbook, this is an easy and clear introduction to the biological and philosophical issues of astrobiology in general.

3. *What are Animals and What are Aliens?*

ANIMAL, VEGETABLE, MINERAL? HOW EIGHTEENTH-CENTURY
SCIENCE DISRUPTED THE NATURAL ORDER BY
SUSANNAH GIBSON

A delightful and easily readable account of the history of classification of life, with lots of examples from weird and wonderful biologist personalities.

MOBY-DICK BY HERMAN MELVILLE

This classic seafaring novel contains a wealth of information about the biology of whales; some, but not all, of which is scientifically accurate.

*THE GARDEN OF EDIACARA: DISCOVERING THE FIRST
COMPLEX LIFE* BY MARK A. S. MCMENAMIN

A slightly technical book, but nonetheless written for a popular audience. If you are interested in geological detective stories, you will probably enjoy this one.

LIFE: AN UNAUTHORISED BIOGRAPHY
BY RICHARD FORTEY

Richard Fortey is a great storyteller, and this account of the evolutionary history of Earth is a good read, and very accessible.

*THE ORIGINS OF LIFE: FROM THE BIRTH OF LIFE TO THE
ORIGIN OF LANGUAGE* BY JOHN MAYNARD SMITH AND
EÖRS SZATHMÁRY

A classic book, although decidedly technical. For those who want more detail, and more rigorous analysis.

LIFE'S SOLUTION BY SIMON CONWAY MORRIS

See Chapter Two.

4. Movement – Scuttling and Gliding Across Space

*RARE EARTH: WHY COMPLEX LIFE IS UNCOMMON
IN THE UNIVERSE* BY PETER D. WARD AND
DONALD BROWNLEE

A slightly pessimistic but ultimately important account of what might make a planet suitable for the evolution of complex life. A good detailed introduction to astrobiology and planetary habitability.

THE COSMIC ZOO: COMPLEX LIFE ON MANY WORLDS BY DIRK SCHULZE-MAKUCH AND WILLIAM BAINS

In opposition to *Rare Earth*, Schulze-Makuch and Bains explore an interesting idea that complex life is, in fact, likely to evolve on other planets.

RESTLESS CREATURES BY MATT WILKINSON

See Chapter Two.

LIFE'S SOLUTION BY SIMON CONWAY MORRIS

See Chapter Two.

WONDERFUL LIFE BY STEPHEN J. GOULD

See Chapter Two.

THE FORMATION OF VEGETABLE MOULD, THROUGH THE ACTION OF WORMS, WITH OBSERVATIONS ON THEIR HABITS BY CHARLES DARWIN

An idiosyncratic but fascinating look at the mind of the great scientist, by means of his detailed notes on the behaviour of earthworms. Available on the internet as facsimile copies.

5. *Communication Channels*

'POSSIBLE WORLDS' BY J. B. S. HALDANE

See Chapter Two.

THE EVOLUTION OF LANGUAGE BY W. TECUMSEH FITCH

A somewhat technical textbook covering the nature and evolution of language across the animal kingdom. Not a light read, but essential for anyone interested in what and why language is.

6. Intelligence (Whatever That Is)

ARE WE SMART ENOUGH TO KNOW HOW SMART ANIMALS ARE? BY FRANS DE WAAL

One of the go-to books on animal consciousness, written by a very lucid and compelling author. This book challenges almost all of the preconceptions that we have about the differences between humans and animals.

WHAT IT'S LIKE TO BE A DOG: AND OTHER ADVENTURES IN ANIMAL NEUROSCIENCE BY GREGORY BERNS

The story of a neuroscientist who put dogs in an MRI machine to see what they were thinking.

ARE DOLPHINS REALLY SMART? THE MAMMAL BEHIND THE MYTH BY JUSTIN GREGG

A comprehensive account of that popular but very misunderstood animal, and those aspects of dolphin life behind the appearance of a smile.

KINDS OF MINDS: TOWARDS AN UNDERSTANDING OF CONSCIOUSNESS BY DANIEL C. DENNETT

Quite a challenging read in many ways, because of its rigorous philosophical approach to consciousness. However, Dennett pulls this off better than many professional philosophers, and the book remains readable and accessible.

MORTAL QUESTIONS BY THOMAS NAGEL

The contrary view to Dennett, but quite a difficult read. For those interested in the detailed philosophical arguments about the nature of consciousness.

THE MISMEASURE OF MAN BY STEPHEN J. GOULD

Gould wrote this book as a condemnation of the tendency to ascribe 'scientific' relevance to claims about differences in intelligence between races. Fascinating reading, and an important reminder of how scientific knowledge can be abused.

THROUGH A WINDOW: MY THIRTY YEARS WITH THE CHIMPANZEES OF GOMBE BY JANE GOODALL

A beautiful account of the research of the most famous primatologist of all time, and her subjects, the chimpanzees she studied to bring us a better understanding of ourselves.

THE GENIUS OF BIRDS BY JENNIFER ACKERMAN

A light-hearted and cheerful account of bird behaviour, and particularly bird intelligence.

CONTACT BY CARL SAGAN

One of the greatest science fiction novels of all time, the famous astronomer takes us through what it would be like to receive a message from an alien intelligence.

THE IMPACT OF DISCOVERING LIFE BEYOND EARTH EDITED BY STEVEN J. DICK

See Chapter One.

THE ALEX STUDIES: COGNITIVE AND COMMUNICATIVE ABILITIES OF GREY PARROTS BY IRENE PEPPERBERG

The definitive account of Pepperberg's work on the linguistic abilities of African grey parrots.

THE BLACK CLOUD BY FRED HOYLE

See Chapter Two.

CALCULATING GOD BY ROBERT J. SAWYER

Hard science fiction at its best, *Calculating God* is more of a philosophical exercise than a shoot-'em-up thriller. An atheist palaeontologist with cancer has his personal belief system challenged when aliens arrive on Earth, declaring themselves to be on a mission to gather evidence for the existence of God.

7. Sociality – Cooperation, Competition and Teatime

THE SELFISH GENE BY RICHARD DAWKINS

See Chapter One.

THE PRIVATE LIFE OF THE RABBIT BY
RON M. LOCKLEY

A beautiful and engaging description of the naturalist's study of wild rabbit behaviour in specially designed enclosures on the country estate of Orielton. The unparalleled insight into the social behaviour of rabbits greatly affected the author Richard Adams, who cites it as the main source for his characters in *Watership Down*.

EVOLUTION AND THE THEORY OF GAMES
BY JOHN MAYNARD SMITH

The standard textbook on the importance of game theory in evolution. Not a simple read, but a resource for those who want to know the details to which I refer only in passing.

ARE DOLPHINS REALLY SMART? BY JUSTIN GREGG

See Chapter Six.

BABOON METAPHYSICS: THE EVOLUTION OF A SOCIAL MIND BY DOROTHY L. CHENEY AND ROBERT M. SEYFARTH

Among the pioneers of primate behaviour research, Cheney and Seyfarth have written a book that brings the world of baboons' social life into vivid perspective.

8. Information – A Very Ancient Commodity

THE COSMIC ZOO BY DIRK SCHULZE-MAKUCH AND WILLIAM BAINS

See Chapter Four.

THE LANGUAGE INSTINCT: HOW THE MIND CREATES LANGUAGE BY STEVEN PINKER

Pinker robustly presents one of the two main streams in evolution of language research. An entertaining and useful read.

9. Language – The Unique Skill

THE LANGUAGE INSTINCT BY STEVEN PINKER

See Chapter Eight.

THE LANGUAGE MYTH: WHY LANGUAGE IS NOT AN INSTINCT BY VYVYAN EVANS

For a contrast to Pinker's position in *The Language Instinct*, this is an amusing read, from an idiosyncratic author.

THE EVOLUTION OF LANGUAGE BY W. TECUMSEH FITCH

See Chapter Five.

THE DIVERSITY OF LIFE BY E. O. WILSON

An evocative overview of ecology, diversity and the mechanisms of how life evolves and coexists on this planet, by a celebrated science communicator.

XENOLINGUISTICS: TOWARD A SCIENCE OF EXTRATERRESTRIAL LANGUAGE EDITED BY DOUGLAS VAKOCH

An edited collection of papers on different approaches to recognizing, and possibly interpreting, alien languages.

10. *Artificial Intelligence – A Universe Full of Bots*

CRABS ON THE ISLAND BY ANATOLY DNEPROV

A wonderfully idiosyncratic Soviet-era science fiction short story about runaway self-replicating robots.

ENDLESS FORMS MOST BEAUTIFUL: THE NEW SCIENCE OF EVO DEVO AND THE MAKING OF THE ANIMAL KINGDOM BY SEAN B. CARROLL

A very influential book on how genes control the form of organisms through their development.

AT HOME IN THE UNIVERSE: THE SEARCH FOR THE LAWS OF SELF-ORGANIZATION AND COMPLEXITY BY STUART KAUFFMAN

This book goes into a certain level of mathematics, but only enough to appreciate the deep elegance of the structures that can evolve spontaneously in simple systems.

ARTIFICIAL LIFE: THE QUEST FOR A NEW CREATION
BY STEVEN LEVY

A relatively easy introduction to the (what was then, in 1992) new field of artificial life, in the sense of self-replicating computer agents (not necessarily physical robots, although Levy touches on this too).

THE MEME MACHINE BY SUSAN BLACKMORE

A good introduction to how the concept of 'memes' works in the world, and how the evolution of ideas can parallel the evolution of organisms.

WONDERFUL LIFE BY STEPHEN J. GOULD

See Chapter Two.

LIFE'S SOLUTION BY SIMON CONWAY MORRIS

See Chapter Two.

SUPERINTELLIGENCE: PATHS, DANGERS, STRATEGIES BY NICK BOSTROM

A very comprehensive (and rather detailed) book on the possible dangers of runaway AI. This book takes perhaps the more pessimistic perspective, but does so in a convincing way.

THE CONSCIOUS MIND: IN SEARCH OF A FUNDAMENTAL THEORY BY DAVID J. CHALMERS
CONSCIOUSNESS EXPLAINED BY DANIEL C. DENNETT

Together, these two authors battle out the question of whether 'mind' is separate from 'body'. Be warned: neither is a particularly easy book to follow, but if you want to investigate this most important of questions, most of what you need will be found here.

11. *Humanity, As We Know It*

LIFE CONCEPTS FROM ARISTOTLE TO DARWIN: ON *VEGETABLE SOULS* BY LUCAS JOHN MIX

Do vegetables have souls? These are the kinds of questions we need to ask when considering the possibilities of life on other planets. It turns out that philosophers from Aristotle onwards considered this question very seriously.

WILD ANIMALS I HAVE KNOWN BY ERNEST THOMPSON SETON

Seton's books, especially this one, are a fantastic portrait of the lives and behaviour of animals, presented in a somewhat anthropomorphic way, but by a skilled naturalist and observer of behaviour.

THE IMPACT OF DISCOVERING LIFE BEYOND EARTH EDITED BY STEVEN J. DICK

See Chapter One.

THE BLANK SLATE: THE MODERN DENIAL OF HUMAN NATURE BY STEVEN PINKER

Pinker presents with conviction his arguments that human nature is a physical biological phenomenon.

SOCIOBIOLOGY: THE NEW SYNTHESIS BY E. O. WILSON

Wilson's take on a similar idea: that human behaviour is fundamentally defined by biology.

ULTRASOCIETY: HOW 10,000 YEARS OF WAR MADE HUMANS THE GREATEST COOPERATORS ON EARTH BY PETER TURCHIN

A description of the role that conflict may have played in shaping our cooperative behaviour.

THE *MAJOR TRANSITIONS IN EVOLUTION* BY
JOHN MAYNARD SMITH AND EÖRS SZATHMÁRY

This is undoubtedly a highly technical book, but if you want the details of the kinds of phenomenal innovations that happened during the evolution of life on Earth, this is the place to look.

Index

Page references in *italics* indicate images.